Toward a Critical Theory of Nature

Critical Theory and the Critique of Society Series

In a time marked by crises and the rise of right-wing authoritarian populism, **Critical Theory and the Critique of Society** intends to renew the critical theory of capitalist society exemplified by the Frankfurt School and critical Marxism's critiques of social domination, authoritarianism, and social regression by expounding the development of such a notion of critical theory, from its founding thinkers, through its subterranean and parallel strands of development, to its contemporary formulations.

Series editors: **Werner Bonefield**, University of York, UK and **Chris O'Kane**, John Jay College of Criminal Justice, City University of New York, USA

Editorial Board:

Bev Best, Sociology, Concordia University
John Abromeit, History, SUNY, Buffalo State, USA
Samir Gandesha, Humanities, Simon Fraser University
Christian Lotz, Philosophy, Michigan State University
Patrick Murray, Philosophy, Creighton University
José Antonio Zamora Zaragoza, Philosophy, Spain
Dirk Braunstein, Institute of Social Research, Frankfurt
Matthias Rothe, German, University of Minnesota
Marina Vishmidt, Cultural Studies, Goldsmiths University
Verena Erlenbusch, Philosophy, University of Memphis
Elena Louisa Lange, Japanese Studies/Philology and Philosophy, University of Zurich
Marcel Stoetzler, Sociology, University of Bangor
Moishe Postone†, History, University of Chicago
Matthias Nilges, Literature, St Xavier University

Available titles:
Right-wing Culture in Contemporary Capitalism, Mathias Nilges
Adorno and Neoliberalism, Charles Andrew Prusik
Spectacular Logic in Hegel and Debord, Eric-John Russell (forthcoming)

Toward a Critical Theory of Nature

Capital, Ecology, and Dialectics

Carl Cassegård

BLOOMSBURY ACADEMIC
LONDON • NEW YORK • OXFORD • NEW DELHI • SYDNEY

BLOOMSBURY ACADEMIC
Bloomsbury Publishing Plc
50 Bedford Square, London, WC1B 3DP, UK
1385 Broadway, New York, NY 10018, USA
29 Earlsfort Terrace, Dublin 2, Ireland

BLOOMSBURY, BLOOMSBURY ACADEMIC and the Diana logo are trademarks of Bloomsbury Publishing Plc

First published in Great Britain 2021
This paperback edition published in 2022

Copyright © Carl Cassegård, 2021

Carl Cassegård has asserted his right under the Copyright, Designs and Patents Act, 1988, to be identified as Author of this work.

For legal purposes the Preface on pp. vi–viii constitute an extension of this copyright page.

Series design by Ben Anslow

All rights reserved. No part of this publication may be reproduced or transmitted in any form or by any means, electronic or mechanical, including photocopying, recording, or any information storage or retrieval system, without prior permission in writing from the publishers.

Bloomsbury Publishing Plc does not have any control over, or responsibility for, any third-party websites referred to or in this book. All internet addresses given in this book were correct at the time of going to press. The author and publisher regret any inconvenience caused if addresses have changed or sites have ceased to exist, but can accept no responsibility for any such changes.

A catalogue record for this book is available from the British Library.

A catalog record for this book is available from the Library of Congress.

ISBN:	HB:	978-1-3501-5950-1
	PB:	978-1-3502-1399-9
	ePDF:	978-1-3501-7626-3
	eBook:	978-1-3501-7627-0

Series: Critical Theory and the Critique of Society

Typeset by Integra Software Services Pvt. Ltd.

To find out more about our authors and books visit www.bloomsbury.com and sign up for our newsletters.

Contents

Preface		vi
1	Introduction: What Is a Critical Theory of Nature?	1
2	Marx's Three Materialisms	27
3	Natural History and the Primacy of the Object	49
4	Capitalism and the Domination of Nature	79
5	Marx, Value, and Nature	99
6	Constellations and Natural Science	117
7	Eco-Marxism's Return to Marx	137
8	World-Ecology and the Persistence of Non-Cartesian Dualism	153
9	New Materialism and Dark Ecology	169
10	Utopia, the Apocalypse, and Praxis	187
Notes		206
References		224
Index		242

Preface

The word "nature" is apt to evoke an imagery of greenery, lush forests, and animals. But nature also means something else. Try imagining your everyday environment—your home, for instance, or the city or village where you spend most of your life. Imagine all the familiar trappings of that environment. Some would call that nature as well, a second nature perhaps. But there is also a third sense of the word that is perhaps especially relevant in today's planetary ecological crisis. Imagine the greenery and forests disappearing. Imagine what life would be without your everyday surroundings. In the threats to our accustomed existence, there is also something that we might call nature. Nature in this third sense stands for the objective forces that undermine the forms through which we experience the world and to which we have grown used. Nature is the hail that hammers on our windows, the pollution that undermines the livability of our planet, the global warming that we seem utterly unable to stop. As these examples show, nature can to a large extent be humanly produced.

In addition to nature, this book also focuses on capitalism, or rather capitalism's production of disasters. The first decades of the millennium have seen an avalanche of catastrophic events, each one hitting from a new direction: financial crises, civil wars, sociopolitical unrest, nuclear meltdowns, pandemics, and increasingly frequent extreme weather that has already become the new normal. Yet despite their destabilizing effects, the capitalist system has shown itself remarkably resilient. In no area is this clearer than in that of climate change. Despite almost complete certitude about the global ecological devastation that will occur if we go on as usual, and despite decades of protest, the system has hardly changed at all. As Alexander Stoner and Andony Melathopoulos point out, today's Great Acceleration coincides "with a world devoid of adequate political practice, in which revolutionary social theory exists without revolutionary practices" (Stoner & Melathopoulos 2015: 74).

The result is a situation in which the unprecedented might of catastrophes is matched by the undiminished stubbornness of the system. Yet, the role of critique is missing from this picture. Critique is the third focus of this book. In the face of the madness of the infernal dance in which the catastrophes and the system reinforce each other in a spiral headed for planetary devastation, critique is our best chance of sanity. While the power of critique might seem slight, let us toy for a moment with the hope that something might grow out of it—a realization, perhaps, that will make us pause, reflect, break with the old, and start to do things differently? Or a realization that action is possible, even when catastrophes and the system conspire to instill a sense of paralysis? A realization that changes can happen unexpectedly, despite the system's stubbornness, and that what matters in such moments is alertness, quick reflection, and a readiness to act so as to influence the direction of that change? And last but not least, mounting discomfort for the powers that be? All of that, I hope, might grow out of critique. If critical theory is to be of use, it must find ways of strengthening this power of critique and thereby tie together knowledge and action. A system change is needed. If this book can contribute to that, I will be pleased.

Many thanks to Chris O'Kane, without whose encouraging comments to an article I wrote a few years back this book would never have been written. Many thanks also to Karolina Enquist-Källgren, my best and most thorough critic, for the time you put into reading my chapters and for your sharp and helpful criticism. One day, I hope, I will be able to convince you! Let me quote a line from your own book, as a homage and small token of gratitude: "Destruction is the proof that neither time nor space can be reduced to a human transcendent structure. For why would the human being, the creator of everything, destroy its own creature?" (Enquist Källgren 2019: 99). Great thanks also to Johan Söderberg, my longtime discussion partner. Like Brecht you remind me that the oppressors do not work in the same way in every epoch. Many ideas that I present in this book have been worked out during our conversations, not least during the doctoral course that we gave together a few years ago on materialism and critique. I'm also grateful for the comments you gave me as my discussant during a seminar presentation at my department. Thank you, Per Månson, for a thorough reading of one of the chapters and for prompting me to articulate clearer where I was going with my arguments.

Thank you, Håkan Thörn, for good cheer and good comments. I wish I could have written more about Marcuse, but I'll leave that book for you to write. I'm also grateful for valuable comments at various stages of the preparation of the text by Mikael Carleheden, Adam Netzén, Anders Ramsay, Linda Soneryd, Alexander Stoner, Carl Wilén, and many others. You have all contributed to making the book better, although I gladly accept responsibility for all faults. I also want to express my appreciation to the participants in the reading groups on Marx, Hegel, and Kant in which I have participated during the last few years and which in various ways have stimulated the ideas that have gone into this book. Reading groups are one of the best inventions of humankind. Finally, I wish to thank Misuzu, Dan, and Lina for your patience, for bringing love and music into my life, and for being wonderful.

1

Introduction: What Is a Critical Theory of Nature?

This book sets out to develop a critical theory of nature. By that I mean a theoretical approach that aims at a critique of the reified social forms that regulate our interaction with nature. I will do my best to unpack this formulation in this chapter by clarifying what I mean by key terms such as "critique," "reified forms," and "nature." But rather than diving straight into the explanatory work, I will start in a slightly roundabout way with two theoretical vignettes that will be useful points of reference for the subsequent discussion.

The first vignette takes us back to Ernst Bloch, great philosopher of Utopia and fellow traveler of the Frankfurt School. In an essay from 1929, "The Anxiety of the Engineer," he paints the following picture of "the Americanized big city" where technology has achieved an apparent victory over nature. The stunning achievements of modern science, he claims, are accompanied by a feeling of vertigo.

> The city of ever-increasing artificiality, in its detachment and distance from the natural landscape, is simultaneously so complex and so vulnerable that it is increasingly threatened by accidents to the same extent that it has rooted itself in midair – that is, the city is built upon roots that have grown more and more synthetic. This grandly suspended, inorganic metropolis must defend itself daily, hourly, against the elements as though against an enemy invasion.
>
> (Bloch 1998: 307)

Today, when the cracks in the gigantic system that we inhabit are increasingly visible, we can surely recognize ourselves in this stunning image of pervasive anxiety. Not only do we seem unable to break out of the pathway to a hothouse earth, we also face a future with uncertain energy supply, an ongoing mass extinction and an increasing vulnerability to starvation and diseases. Among many, the anxiety is already giving way to panic and depression. To Bloch's picture we might add that today's environmental anxiety is exacerbated by the fact that many detrimental environmental effects only come with a time lag. Not only do we fear that things may go horribly wrong at the slightest misstep, we also fear that they may already be beyond repair. Like Oedipus, we fear catastrophes that we may *already* have caused, without being fully aware of it. In that sense, the anxiety comes with what Freud calls the uncanny, the return of familiar things that have been repressed. As Tim Morton (2013: 99) remarks, pleasant conversations about the weather have become impossible. Sunny days remind us of global warming and give us bad conscience. The city may imagine itself as rooted in midair, but what is repressed and shows up as anxiety is our awareness that it is in fact rooted in a soil that might be more than inert, obedient matter available for control. Friedrich Engels put it well in a classic passage in which he urges us not to flatter ourselves on account of our human conquest over nature: "For each such conquest takes its revenge on us" (Engels 1987b: 460). As in Bloch's imagined city, we fear the punishment that awaits us for our technological hubris.

The second vignette takes us further back in time, to Hegel's *Science of Logic*, written at the cusp of the industrial revolution in Europe The section on which I want to dwell is probably the clearest expression in his writings of the famous dialectical leap, or *Umschlag*, from quantity into quality—or to use his own expression: "the sudden conversion into a change of quality of a change which was apparently merely quantitative" (Hegel 1969: 335). According to Hegel, everything that exists is determined by magnitudes, or quantities. If a dwarf grows beyond a certain size it is no longer a dwarf. A piece of barren land must exceed a certain size before we call it a desert. To take a contemporary example, hot days must continue with a certain regularity before we speak of climate change, and so on. The size or magnitude (a quantity) is thus part of what defines the thing (a quality). The specific quantity, or quantum, is "the determination of the thing, which is destroyed if it is increased or diminished beyond this quantum" (Hegel 1969: 333f).

Hegel's term for the magnitude that defines an entity is "measure." Measure is paradoxical. This is because we usually don't find anything in the concept of a thing that pinpoints the exact quantitative limit where a qualitative change must occur. This makes us think that we can vary the quantity without affecting the quality. A forest remains a forest even if we cut down one tree. A heap remains a heap even if we remove a grain of sand from it, and a hair pulled from a person's head won't make the person bald. But, obviously, if we keep cutting, removing, and pulling we eventually arrive at a point where the forest and the heap disappear and the person turns bald. What Hegel describes is the so-called *sorites* paradox (from the Greek word for heap). He concludes that

> the destruction of anything which has a measure takes place through the alteration of its quantum. On the one hand, this destruction appears as *unexpected*, in so far as the quantum can be changed without altering the measure and the quality of the thing; but on the other hand, it is made into something quite easy to understand through the idea of *gradualness*.
>
> (Hegel 1969: 334f)

In a magnificent passage that comes like a bolt out of the blue in Hegel's otherwise seemingly apolitical discussion, he mentions the destruction of the State or of great fortunes as two examples:

> Quantum ... is the aspect of an existence which leaves it open to unsuspected attack and destruction. It is the cunning of the Notion to seize on this aspect of a reality where its quality does not seem to come into play and such is its cunning that the aggrandizement of a State or of a fortune, etc., which leads finally to disaster for the State or for the owner, even appear at first to be their good fortune.
>
> (Hegel 1969: 336)

We should note that Hegel points out that the destruction is always *unexpected*, and this *despite* it being easy to understand what causes it. Curiously, it's not ignorance that makes us surprised but our knowledge. We are surprised because our concepts tell us that small quantitative changes in things won't affect their quality. We know that removing one grain of sand from a heap won't make the heap disappear, because that's part of the concept of a heap. That's why politicians and capitalists, and others too, are justified in thinking that a little more aggrandizement won't hurt.

Who can avoid thinking here of global warming? Even though we are aware that releasing greenhouse gases into the air will cause global catastrophe, we go on thinking that a little more won't hurt. And we are quite correct in thinking so. Surely, it can't possibly make any difference to the global climate whether we release an additional CO_2 molecule into the air or not, or whether we take the car to work or not. It's precisely because we are correct that we will be so *surprised* when the catastrophe arrives. To put it differently, we can't stand with a measurement instrument in hand and say: *now* the catastrophe begins. What happens is rather that, when the dialectical *Umschlag* finally occurs, we realize that the catastrophe has been going on all the time, and that we were in its midst even when things still seemed fine. The shift of perspective surprises us since it projects the origins of the catastrophic process back into the past. To awaken to the catastrophe of global warming is to realize not only that the catastrophe is here, but that it has been unfolding for a long time without us being aware of it, perhaps ever since the first steam engines. The dialectical transformation that Hegel describes is not just a one-way process whereby quantitative changes give rise to qualitative ones. It also involves an opposite movement through which the qualitative shift produces a new version of the past which points out which quantitative processes are relevant and important. That's why even a sudden awareness of catastrophe often includes the realization that the catastrophe is old. The catastrophe started already when we cut the first tree, removed the first grain of sand, or plucked the first hair.

Hegel tells us that catastrophes can be understood only if we focus on both quantitative and qualitative processes. Catastrophes consist, on the one hand, in complex, interlocking processes of a quantitative nature, but, on the other hand, they are also conceptual *Gestalt* shifts, or shifts in perspective, that make us see things in a qualitatively new manner. We are compelled to these qualitative shifts by quantitative processes—such as the increase of atmospheric CO_2—but the compulsion is not of a causal nature, in the sense of a law-like regularity or automatic reflection. The *Umschlag* is not mechanical. It is a response to surprise, just like when we suddenly realize that a heap of sand is disappearing. This shows that Hegel, despite being an idealist, does not disregard experience. Concepts develop in response to experiences, and this development would have been unintelligible without the latter.

As for now, let us just remember that quantitative changes can be subversive. That insight is a source of anxiety but it also provides an opening for idealist self-criticism. It is hard to grasp the brittleness of things if we focus only on their quality, on the way they appear to us through our concepts. No matter how correct and justified we are in clinging to the quality of the world as we comprehend it, quantitative changes continually undermine and destabilize it, precisely because they take place below the radar of conceptual thinking.

Critical Materialism, Second Nature, and Catastrophe

What do these two vignettes tell us about a critical theory of nature?

Hegel and Bloch show how our attempts to master an unruly reality generate both anxiety and surprise. This relation, I suggest, can be grasped as a relation between *form* and *matter*. Form consists in the principles that structure our relationship to matter while matter has the potential to undermine form. This relation is central to the critical theory of nature.[1] The book will explore it in a variety of guises, for instance, when we discuss what Theodor W. Adorno calls the non-identity between concept and object or when we discuss Marxian concepts such as value and use-value or reification and metabolism. Exploring this relationship will help us understand both why forms seem so impervious to change and why they're so brittle. As the two vignettes show, the semblance of stability can coexist with gradual processes and incongruities that, we sense, may develop into sweeping transformations.

That forms can be undermined by matter is relevant not only for understanding how society relates to nature but also to the possibility of critique. Both Bloch and Hegel point to the methodological importance of feelings such as anxiety or surprise for breaking out of the seeming closure or self-sufficiency of forms. Such feelings become a source of critique since, no matter how much society might appear rooted in midair, we sense that this semblance is false and that it remains dependent on elements that it shuts out from view. Marx too focused on the relation between form and matter as the site of contradictions that could be mobilized for the sake of critique. A clear example is the attention that he pays in *Capital* to experiences of the working day as glimpsed in reports and newspapers (Marx 1990: 340–416). The point of this material is not simply to flesh out his theory with empirical detail, but also to illuminate the contradiction between the dominant economic categories and the experience of the workers. It is inserted with a critical intent, as part of the critique of political economy.

This procedure for bringing out contradictions also characterizes the critical theory of nature and defines this approach as *critical* rather than *traditional*, to use Max Horkheimer's classical distinction (2002a). While traditional theory aims at a neutral description of objective processes in nature or society from the standpoint of a contemplative observer, critical theory seeks to bring out contradictions, sharpen our awareness of them, and thereby strengthen the opposition to oppression and exploitation. Critical theory is supposed to be practical and emancipatory. In relation to nature, this means that its task is not primarily to conduct scientific inquiries into nature, but to sharpen our awareness of how we, as subjects, are related to nature and how we may relate to nature in our praxis.

Bloch's and Hegel's mention of cities, states, and great fortunes underscores that criticizing forms is no mere intellectual exercise. Forms are not just mental constructs but also a material reality. Importantly, the forms constitutive of the capitalist economy are reproduced as so-called "real abstractions" (Sohn-Rethel 1978) as part of people's economic behavior. The value-form, for instance, is reproduced whenever we exchange products on the market whether we are conscious of it or not. As Marx writes, it is reproduced "behind the backs" of individuals since it is part of how capitalism works (Marx 1990: 135). Bloch is lucid in illuminating how the forms through which we relate to nature express the persistence of class and capitalism. It is that persistence, rather than any inherent property in technology per se, that has produced the present situation in which technology "stands in nature like an army of occupation in enemy territory" (Bloch 1995: 696). As we will see, the forms targeted by the critical theory of nature include not only the value-form but all forms that are systematically put into play and reproduced as part of capitalist society's interaction with nature, including the division between "society" and "nature."[2]

In the sections to follow I lay out the outlines of the critical theory of nature. I begin by taking up the concept of the Anthropocene to indicate the contemporary relevance of the dialectics of form and matter. We are then ready to have a closer look at the role assigned to matter in criticizing reified forms and how this role is connected to a distinctive form of materialism, which I call critical materialism. Three terms will play an especially important role: capitalism as second nature, materialism as critical practice, and catastrophes as the language of objects. A critical theory of nature needs, I suggest, to keep all three in view. Having argued for this threefold focus, I move on to explain the book's usage of the concept of nature and a final section in which I sketch the overall structure of the book.

The Anthropocene, Second Nature, and the Reification of Nature

Much of the anxiety that Bloch discussed in connection with the "city of ever-increasing artificiality" is condensed and heightened to an unprecedented degree in the notion of the Anthropocene, which in recent years has gained extraordinary popularity as a term for the present geological epoch. Why, if it stands for the age in which humankind affects the earth in the manner of a geological force, does it seem so indelibly marked by anxiety, and even panic and regret? If it means that humanity is in charge, why does it seem to spell doom rather than freedom? How do we account for the paradox of the Anthropocene, which the environmental philosopher Clive Hamilton has referred to as the "bizarre situation, in which we have become potent enough to change the course of the earth yet seem unable to regulate ourselves" (2017: viif)? We can rephrase this by asking: why do the social *forms* that structure our relationship to nature appear to be so resistant to change, despite an overwhelming number of people being aware of the catastrophic consequences of not changing them?

The paradox disappears when we recall that the immense powers referred to in the notion of the Anthropocene are not humanity's but those of a second nature, namely, capitalism. As Daniel Cunha (2015: 68) points out: "That Man is presented as a blind geologic force, such as volcanic eruptions or variations in solar radiation, is an expression of the naturalized or fetishized form of social relations that is prevalent in capitalism." According to Marx (1993: 196), capitalism generates a semblance of naturalness despite being a historical creation. Its laws come forward as ahistorical and eternal, as a fate that humans are unable to change. Georg Lukács (1971: 128), in whose writings the term "second nature" was applied to capitalism for the first time, defines it as a semblance, arising in society itself, of law-like and objective regularities beyond human control.[3] In a sentence that strikingly anticipates the Anthropocene, Horkheimer and Adorno point to the "panic fear" that accompanies the perception of society as a second nature: "men expect that the world ... will be set on fire by a totality which they themselves are and over which they have no control" (Adorno & Horkheimer 1997: 29). The fear inspired by the Anthropocene is a fear of our ghostlike fetishistic double, the second nature generated by capitalism that now confronts us as a deadly force threatening all life with extinction.

Analyzing and criticizing the grip of second nature on us and thereby helping us resist it is perhaps the central and most urgent task of a critical theory of nature. Here a word should be said about reification, the process through which the world comes to appear as a second nature. According to popular understanding, reification means treating human beings or human creations as things, but this is not how the concept will be used in this book.[4] I believe that this popular understanding is at odds with or at least overlooks significant aspects of how the concept was used by the generation of social thinkers who introduced it in social thought—thinkers like Lukács, Bloch, and Adorno. To them reification had little to do with the opposition of humans and things, but rather meant that an object, human or not, appeared to possess an essence or substance independently of the process of its historical mediation. In this sense, even the act of identifying a person as human in distinction to, say, animals or things can be an instance of reification. Reification is not the opposite of the human but of the historical. "For all reification is a forgetting: objects become purely thing-like the moment they are retained for us without the continued presence of their other aspects: when something of them has been forgotten," Adorno writes in a letter to Walter Benjamin (Adorno & Benjamin 1999: 321). What Adorno describes here is how our perception of the world becomes colored by a reifying thing-form (or form of objectivity, *Gegenständlichkeitsform*, as Lukács called it). To me, this operation is the kernel of the notion of reification. Reification, in other words, is not just to treat human beings or human creations as things, but to treat *any* aspect of the dialectically constituted world as a thing.

In the usage of Lukács, the roots linking the concept of reification to Marx's idea of commodity fetishism are clear. Just as the commodity appears to possess value in its own right, independently of the process of production, the reified "thing" appears to possess an essence in its own right, regardless of social or historical context. Theoretically, the idea of second nature helps us understand why the social forms of capitalism appear so resistant to change. Second nature isn't just a subjective misrecognition or inability to perceive historical creations as changeable. It is rooted in the real abstractions constituted by the exchange relations through which capitalist society is reproduced. Here we see part of the answer to the riddle of the helplessness that we seem to feel in the face of the Anthropocene.

Unfortunately, there is a small glitch in the notion of reification that can easily turn into a hidden trap. The interpretation that the second nature is artificial while the first is not invites the reproach that the argument about reification rests on a questionable ontological separation between nature and society. Lukács himself primarily applies the concept of reification to how *social relations* are transformed by capitalism and modern bureaucracy. However, as critics have pointed out, restricting the critique of reified thinking to social relations implies that reified thinking is legitimate when applied to nature. Jane Bennett (2010: xiv), for instance, rejects such "demystifying critique" since she believes that it aims only at uncovering human agency while viewing nature as naturally thinglike.

To ward off the misunderstanding that the concept of reification is anthropocentric, it should be clearly stated that *non-human nature too can be reified*. Nature, like society, is reified when the thing-form is imposed on it, which happens, for instance, when it is treated solely under the aspects that happen to be relevant to markets, bureaucracies, or other dominant institutions in society. The incorporation of nature into the dominant systems of society—as an industrial resource, an item for consumption, or an object to be administered—is almost invariably accompanied by its reification. Extending the notion of reification to nature appears to me to be perfectly justifiable from the point of view of Marx's theory of capitalism: the real abstractions that are reproduced by capitalism do not just impose forms functional for the system on social relations, but also forms for processing nature, both as a source of use-values and as a commodity.

If nature can be reified, then it follows that the distinction between first and second nature doesn't concern the *content* of the two concepts so much as the *manner* in which this content is approached. What matters is not whether the content is artificial or not, but whether it is reduced to aspects relevant for dominant institutions or not. With this interpretation there is no ontological separation between society and nature. Instead of starting with the separation and then arguing that reification is illegitimate when applied to society but not in relation to nature, reification itself becomes a criterion for distinguishing between first and second nature. Second nature differs from first nature, not by being artificial but by being reified. A forest that is subjected to the market is second nature, not first nature. The same holds for animals that are regarded as nothing but resources or merchandise. The subjection of a forest or an animal to market imperatives implies reification no less than the subjection of the working class.

While the concepts of reification and second nature are indispensable to a critical theory of nature, we should take care to avoid interpreting them anthropocentrically. Reification doesn't mean treating humans as things, nor does dereification mean revealing the human origin of what appears natural. Such an interpretation would pull the rug from under the concept of a reification of nature. Reification means treating *anything*—whether human or not—as a thing that can be reduced to the aspects that dominant institutions like markets or bureaucracies define as relevant. A world constituted by such things is second nature, whether it is artificial or not.

Critical Materialism

We can now specify the contours of the critical theory of nature more clearly in relation to other perspectives.

In developing a critical theory of nature, I draw on the tradition from Marx to the Frankfurt School. Nature has been a prominent connecting thread running through this tradition, as indicated by Marx's insights into capitalism's destructive metabolic relation with nature, by the critiques of instrumental reason and the domination of nature in classical works like Horkheimer and Adorno's *Dialectic of Enlightenment* and Alfred Schmidt's *The Concept of Nature in Marx*, and by the speculations about a reconciliation with nature, or even its liberation, in the writings of early Marx, Bloch, Benjamin, Adorno, and Herbert Marcuse. Recent decades have seen a rising number of scholars who are inspired by this tradition to conduct critical analyses of the relationship between modern society, technology, and nature in addition to many valuable and illuminating works on the concepts of nature and materialism in Marx, the Frankfurt School and Western Marxism generally.[5]

This book, however, is not a history of ideas and, rather than trying to do justice to this entire tradition, it seeks to extract key ideas from it and to demonstrate how they can be made fruitful for understanding today's planetary environmental crisis.[6] In particular I will make recurrent use of ideas associated with Adorno—ideas expressed by terms such as "catastrophe," "reconciliation with nature," "experiences of non-identity," and "constellations." I want to suggest that the best way to see how these ideas hang together is to view them from the point of view of what I call *critical* materialism. I will return to this materialism in the next chapter, but already here I want to indicate briefly how it differs from two other prominent forms of materialism in the Marxist tradition.

To begin with, critical materialism differs from the *causal* materialism associated with so-called scientific Marxism and Engels's dialectics of nature. The core idea here consists in the central role assigned to matter as a causally determining factor, most famously expressed in the base-superstructure metaphor. The model for this materialism is, implicitly or explicitly, natural science rather than Hegelian dialectics. Its aspiration is to demonstrate the applicability of scientific methods to history, thus permitting the discovery of historical laws. Since the causal relation between base and superstructure must ultimately be posited as unidirectional if the term "materialism" is to carry any meaning, and since it plays itself out between fixed, reified entities (such as "nature," "the economy," and "ideology"), this approach hardly leaves any room for grasping this relation dialectically, as constituted by mutually determining elements.

This causal form of materialism was attacked by Lukács, Antonio Gramsci, and other so-called Western Marxists, who instead developed what I call a *practical* materialism. In practical materialism it is not the economy conceived of as a base that constitutes the material element, but rather human praxis as mediated through meaningful contexts or "totalities." Causal necessity is replaced by conceptual necessity, since the totality is a retrospective reconstruction of the meaning of life contexts. While this form of dialectics is not able to predict anything scientifically, it can help revolutionary anti-capitalist activists make sense of their struggle, elucidate why negating capitalism is a meaningful activity, and demonstrate how this activity relates to the overall constellation of historical conditions and opportunities. Causal determinism is thus rejected in favor of an orientation to political action. Future developments can only be apprehended by dialectics to the extent that such developments have already emerged and become integrated into how we make sense of our present. Rather than natural science, here it is social science—and perhaps above all studies of social conflicts, movements, resistance, and revolutions—that can be of help by directing attention to ongoing struggles and points of conflict.

Critical materialism differs from the abovementioned varieties of materialism since it employs materialism as a tool for critique, taking as its model Marx's critique of political economy in *Capital*. Rather than presenting its own positive account of history, it focuses on the constitutive relations within the system of capitalism, criticizing the latter immanently by confronting it with its dependency on nature and the exploitation of labor. Criticizing dominant systems takes precedence over constructing conceptual systems of its own. Far from being a base, the capitalist economy is itself seen as an idealist system that imposes its forms on an external reality consisting of both human and non-human nature. Since the system does not express the truth of its relations to this outside, it is a false or *negative* totality. The material element in this materialism is not the economy, but the system's *outside*, the real world on which the system depends. Rather than being a category of the system, matter is the environment on which the system depends but which the system's categories can never do justice. As Hegel highlighted in his discussion about measure, matter can therefore undermine and disrupt the categories that constitute the system. It is as an embodiment of matter that Fredric Jameson evokes history as a force that will make itself known to us whether we care to acknowledge it or not:

> History is what hurts ... This is indeed the ultimate sense in which History as ground and untranslatable horizon needs no particular theoretical justification: we may be sure that its alienating necessities will not forget us, however much we might prefer to ignore them.
>
> (Jameson 1981: 102)

The passage hints at how a notion of reality can be expressed that is viable even to a readership steeped in postmodern skepticism. Reality is not history as an objectified body of facts or of interpretations, but as what destroys our expectations. If it hurts, then it is real, or, as Adorno would say: the true is what does not fit in (Adorno 1997: 59). Even if reality cannot be directly expressed in the terms provided by our conceptual systems, it can be known as what is non-identical to them—as what disrupts, resists, or fails to conform with the systems. Non-identity means that concept and object differ, or in other words that objects are never fully subsumed by their concepts.

This means that critical materialism follows Benjamin in making the sensation of shock central. Shock is like the anxiety and surprise referred to by Bloch and Hegel. It is also like the vertigo, shudder, and pain mentioned by Adorno as instances that allow us to see the untruth of the systems (Adorno 1975: 43, 1997: 331). Like shock, these sensations are negative ways in which consciousness registers the discrepancy between itself and external reality. Under modern conditions of reified thinking, they are the language by which changes on the material level are communicated to the subject. For someone aiming at a materialist investigation, they are both politically and methodologically important since they effect a rupture in the mythically closed dream-world of capitalism through which we can catch a glimpse of materiality itself. They become a means for breaking the hold of reified forms on our thinking. They are therefore appropriate focal points *par excellence* for the critical theory of nature, since they are the negative impact of reality on ideas. Attending to them becomes a way to criticize the system, to expose the distortions it forces on our understanding of history, society, nature, and ourselves.

Permanent Catastrophe

The shocks through which material objects communicate with the subject can be subdued or miniscule, but they also come in the form of large-scale environmental catastrophes such as hurricanes, pandemics, or desertification. The great impact of such catastrophes makes it seem frivolous to compare them to the fleeting shocks documented by Benjamin in connection with city crowds and machinery. Yet their common denominator is that they expose the limits of idealism. Whereas idealism presents history as meaningful, a history punctuated by shocks and disasters is materialist, scarred by the irruption of meaningless events. Such a history comes forward as what Benjamin and

Adorno referred to as a permanent catastrophe (see e.g. Adorno 1973a: 320, Benjamin 1977b: 255, 1999: 473). Today that notion is a fitting lens through which to view the unfolding string of disasters associated with today's incipient mass-extinction, global warming, and the exhaustion of natural resources.[7]

Like the qualitative changes discussed by Hegel, the permanent catastrophe doesn't simply reflect a quantitative increase in disastrous events. It results from a shift of perspective that involves a retrospective movement. It turns "When will the catastrophe begin?" into "When did it start?" By affirming the idea of permanent catastrophe, Benjamin and Adorno hold on to the vision of history that becomes visible in this shift of perspective. "That things 'go on as usual' *is* the catastrophe," Benjamin writes. "It is not what always lies ahead but what is always given. Strindberg's thought: hell is not what awaits us, but *this life here and now*" (Benjamin 1977a: 246). Here we see how they differed from Hegel, who was an idealist due to his belief that mind would ultimately be able to redeem even the slaughter-bench of history by endowing it with meaning (Hegel 2001: 35). To Benjamin and Adorno, by contrast, history contained too much suffering to be justified. It was bound to be forever fragmented, with no other unity than the jarring continuity of discontinuity (Adorno 1975: 314).

Such a view of history also underlies the critical theory of nature. Let me briefly mention four consequences of that. To begin with, a history seen as permanent catastrophe is not deterministic. It tears away confidence in history, but it is open, precisely because it is materialistic. Secondly, the catastrophes give the lie not just to idealist philosophy in general but also to the pretense that the capitalist system is working fine. Even more acutely than the anxiety of Bloch's engineer, catastrophes demonstrate that the city is rooted not in midair but in nature and is therefore vulnerable.

Thirdly, social movements too must abandon idealism. Movements can't transform the world on their own while disregarding the recalcitrance of matter, especially not under deteriorating ecological conditions. Trusting in catastrophes alone to change the system is even worse. In addition to the fact that suffering and devastation often hit the poorest and most vulnerable worst while leaving elites relatively unscathed, catastrophes regularly benefit capital by increasing opportunities for disaster capitalism and "nature washing" (Klein 2008, Smith 2007, 2010: 245ff). For catastrophes to result in qualitatively new and better ways of organizing society, they must be accompanied by critique and mobilization. Challenging the system must happen through an *interplay* between organized movements and the revenge of nature. The catastrophes can help make the critique materialistic but are not a replacement for movements or revolutions.

A final consequence concerns how the critical theory of nature should relate to environmental apocalypticism. There has long been a debate among environmental activists around the pros and cons of "catastrophism," the standpoint that society is headed for collapse. That doomsaying is politically unhelpful, and that it can be paralyzing and detrimental to activism, is a point well taken (Katz 1995, Lilley et al. 2012). That alarmistic warnings about the future have to a large extent become recuperated into the established discourse of green capitalism is also true (Harvey 1996: 148f, 193ff, Methmann & Rothe 2012, Smith 2006, 2010: 245ff). I fully agree with the criticism of a catastrophism purely oriented toward future threats, usually presented as avoidable if only the "right" steps are taken through some mix of market mechanisms and green technology. Such a stance is not only easily coopted by the system, but also disregards the real disasters that are already happening with increasing frequency, in the global south as well as in the global north. In contrast to such catastrophism, it should be emphasized that the apocalypse is not a future event, but *already* here—although it is "combined and uneven," as Evan Calder Williams puts it, progressing in uneven paces and interacting with local circumstances (Williams 2011; compare Swyngedouw 2010, 2013).

That horrible catastrophes are already here, especially for the most vulnerable people and species, is a fact that must be accepted and given the proper attention. Catastrophism is not a matter of strategy. Debating whether it helps mobilization or not presupposes a distance that we no longer have. We shouldn't pour scorn on "left apocalypticism" as some Marxist scholars have done (Smith 2010: 247f). But neither should we rest satisfied with gesturing to future threats, as many environmentalists have done since the start of the modern environmental movement. Instead, a critical theory of nature must attune itself to the catastrophic rhythm that increasingly engulfs the world and accept the ongoing disasters as a *premise* for further political action. Demanding justice, redressing wrongs, halting destruction, and searching out forms of solidarity that don't exclude the weak are tasks that remain as urgent as ever, even in the midst of catastrophes. As I will argue repeatedly throughout this book, accepting the reality of catastrophes does not amount to defeatism. The idea of permanent catastrophe in Benjamin and Adorno is not meant to deprive us of all hope, but to shock us out of the complacent trust in prevailing ideologies. If catastrophes are how nature resists the forms imposed on it, then they can also be mobilized for a critique of the system. Struggles that have lost none of their urgency include demanding and working for a radical transformation of the economy to stop destructive practices. Massive transfers of wealth and power are needed to "democratize survivability" (Hamilton 2010) and encourage at least a semblance of egalitarianism and social solidarity. While this implies a break with the logic of value, of blind capital accumulation for its own sake, the struggle against capital needs to be waged without any naïve trust in the powers of the state, which can easily turn into a "climate leviathan" (Wainwright & Mann 2018) that uses the catastrophes to legitimize oppression. It also needs to be waged without romantically idealizing nature as necessarily nurturing and beneficent, which it certainly isn't after two centuries of resource exploitation and pollution. The immensity of these tasks and the reality of ongoing losses may seem depressing, and it is quite clear that nothing whatsoever guarantees that the struggles will be successful. Indeed, setbacks are part of the permanent catastrophe. To nevertheless wager on the meaningfulness of critique and struggle in this situation implies giving room for emotions that are sometimes considered demobilizing—emotions such as pain, longing, grief, and depression (Weber Nicholsen 2002). We must start from those emotions, not hiding the contradictions but naming them, bringing them out in the open, and think and act from there.

The Concept of Nature and Cartesian Dualism

This book is about the critical theory of nature, but what is *nature*? Is it simply a form imposed on matter, and in that sense a social construction, or is it matter itself, the underlying substance on which form is imposed?

The short answer is that nature is both. As a *form*, capitalist societies reproduce it above all as a sphere of free appropriation to serve the needs of capital. As *matter*, however, it can undermine form. This means that a critical theory of nature adopts a double focus when it approaches nature. As I show in Chapter 3, Adorno works out such an approach through his notions of natural history and the primacy of the object—the former of which traces the shifts in the concept of nature as it gets redefined in relation to history whereas the latter notion allows us to think of nature as more fundamental than the concepts that are used to cover it. Whereas the former notion points to the socially constructed aspects of nature, the latter notion points to nature's irreducible objectivity. By interrelating these two notions, Adorno helps us see the one-sidedness of both constructivism and realism. He thereby helps us respond to the call—voiced, for instance, by Kate Soper (1995: 8) and Andrew Biro (2000: iv)—for an approach to nature that takes seriously the materiality of ecological crises without succumbing a naïve realism.

To expand on this short answer, Raymond Williams (1980: 67–85) is right that nature is one of the most complex concepts in language.[8] Nature is not only a realm of inert, soulless matter, and as such a fitting object for science and technology, but also a romanticized other, standing for what are in the process of losing through industrialization and civilization, a source of a truer way of relating to our own selves. While both these perspectives posit nature as separate from us, nature can also be conceived as including humanity, as in ecological perspectives. Nature can both be conceived of as something to be nurtured, protected, or worshipped, and as an oppressive sign of unfreedom or a semblance of mythic taken-for-grantedness that hides power relations and that needs to be deconstructed—an ambiguity that, as Soper (1995: 4f) points out, is reflected in the co-existence of "nature-embracing" and "nature-skeptical" attitudes in the contemporary movement scene, with environmentalists supporting the former and feminists and anti-racists the latter.

Rather than viewing this surfeit of meanings as a hindrance to the concept's usefulness, I see it as a good illustration of its socially constructed side, the fact that nature must always be understood in its social context. Keeping track of its various problematizing, subversive, utopian, romanticizing, and ideological uses is part of what a critical theory of nature must do. However, that nature has a variable, socially constructed side doesn't contradict its material existence, which we can experience sensorially and which makes itself felt in the fragility of bodies as well as social structures. Claiming that nature lacks a fixed content is not to deny its objectivity. Doing so would be idealist conceit, Bloch's city in midair. Nature is objective because it cannot be reduced entirely to what is subjectively meaningful to us. It is always to some extent beyond the grasp of our concepts, but that doesn't mean that we cannot know it in part. We know it through experiences that upset our concepts and make us recognize their insufficiency, such as the anxiety, surprise, shocks, and pain.

Eco-Marxism and the Production-of-Nature Approach

Against this background, we see clearer how a critical theory of nature differs from other Marxist approaches to nature.[9] Let us take eco-Marxism and the production-of-nature approach, two highly influential perspectives that have sprung up in recent decades and represent contrasting positions in regard to the issue of realism versus constructivism. Both are approaches with which I will attempt a dialogue in several of the chapters to follow.

Eco-Marxism has focused more systematically than other theoretical traditions on demonstrating Marx's usefulness for understanding ecological concerns. John Bellamy Foster's *Marx's Ecology* (2000) and Paul Burkett's *Marx and Nature* (2014) in particular are seminal works highlighting the fruitfulness of Marx's concept of a metabolic rift and of a value-form approach for grasping the relation between nature and capitalism. More than other strands of Marxism, eco-Marxists have emphasized the natural limits to capitalism and human life, claiming that the relation to nature constitutes the primary contradiction in capitalism. Epistemologically, eco-Marxism tends to a naturalist realism, seeing nature as an objective ecological realm constituting a crucial material base for everything else, including human society. A Marxist dialectic should not reject natural science but must find ways of incorporating its methods and theories. Understanding human metabolism with this realm solely by means of a social or cultural analysis is a serious mistake. Not surprisingly, eco-Marxists have sharply criticized constructivist views on nature, both within and without the field of Marxist scholarship.

The production-of-nature approach is more loosely held together. Many adherents are geographers, such as Neil Smith, David Harvey, and Noel Castree. Its distinctive claim is to emphasize capitalism's role in the production of nature—a point most explicitly made in Smith's influential *Uneven Development* (2010). In Smith's view "first" nature has for all practical purposes been subsumed by an artificial "second" nature governed by capitalist exchange. Rather than dominating an external nature, capitalism reproduces nature as an integral part of itself. The production-of-nature approach therefore tends to reject dualist approaches separating nature and society, and, in line with this, to downplay the idea of natural limits. Although recognizing the existence of un-produced parts of nature that are irrelevant from an economic point of view (such as outer space and deep geological strata), it represents a qualified constructivist position regarding nature.[10]

The critical theory of nature is neither as realist nor as constructivist as these two approaches. More than eco-Marxism, it has a clear eye for the fluidity of nature as a social category, tracing how the semblance of nature reappears as a second nature in society itself. At the same time, it is keener than production-of-nature scholars to point to the independent and external existence of a non-human first nature, dominated by the second one. This middle position is not a mere compromise but results directly from its critical procedure in which the dialectics of form and matter is central. Unlike the two other approaches, the critical theory of nature adopts an immanent critique that confronts social forms with the contradictions that are generated when these forms are imposed on matter.[11]

This procedure unavoidably throws a critical light on the *realism* of eco-Marxism. While realism presupposes the correspondence between concepts and reality, the critical theory of nature questions the assumption of such correspondence, which, in Adorno's terminology, disregards what is non-identical in objects. This implies a critical scrutiny of natural science, which is seen as necessarily mediated by conceptual thinking and by the society in which it is situated. It also means that when critical theory speaks of nature, it tries to do so in a way that avoids the risk that its own accounts of nature will lead to new ways to instrumentalize and dominate it or otherwise contribute to its reification. None of this means that critical theory is *constructivist* in the sense of the production-of-nature approach. Without the objective existence of nature, it would be meaningless to say that our categories do violence on it. Constructivism commits the same mistake as realism since it disregards nature's non-identity with our concepts and its capacity to resist. By ruling out the possibility of an incongruence between form and matter from the start, constructivism blunts its critical edge in regard to capitalism's relation to nature—an argument to which I will return in Chapter 8.

Cartesian Dualism

Environmentalists have often blamed environmental destruction on the central place in modern Western culture of a Cartesian dualism that sees the human subject (*res cogitans*) as the sole bearer of rationality and freedom while nature is regarded as nothing but inert, lifeless matter (*res extensa*).[12] This dualism is institutionally anchored in the nature-society divide characteristic of capitalist societies and underpins the view of nature as a mere resource to be freely exploited.

Here I want to make three interrelated points that clarify that the critical theory of nature at no point affirms anything close to Cartesianism. The first point is that the rise of capitalism has historically given rise to a division between society and nature that cannot be wished away. Recognizing the historical context of the emergence of this division doesn't mean that we must stop using the categories of nature or society. We can use them without having to affirm Cartesian dualism as ontologically valid. The crucial point is that critical theorists see the division as a historical creation and seek to abolish its foundation in capitalism, while Cartesian dualists defend it as timelessly true.

The second point develops the first point while modifying it. By introducing a distinction between capitalism and the subject that criticizes it, critical materialism introduces a split in the category of society. It therefore in fact operates with *three* terms: the critical subject stands opposed not just to first nature but also to a second nature represented by capitalism. Unlike in environmentalist arguments about the Cartesian dualism, this model points to capitalism as responsible for environmental destruction rather than humanity in the abstract. While both first and second nature come forward as governed by natural or quasi-natural laws, the subject represents the possibility of agency or, in other words, of a reawakened history. Again, this is fundamentally different from Cartesian dualism since the procedure aims at abolishing the systematic causes of the destructive split between society and nature, and since it doesn't exclude possible alliances with non-human subjects. We can add that this threefold focus of the critical theory of nature also differs from the predominantly *twofold* focus of mainstream Marxism that stresses the class struggle of the bourgeoisie and proletariat. While class struggle is important, an exclusive focus on it leaves the role of nature obscure. An important consequence of introducing the threefold focus sketched here is that the anti-capitalist struggle can no longer be conceived of solely as the struggle of human subjects; instead, as suggested above, it must be conducted through an *interplay* between critical movements and natural forces.

The third point returns us to the relation between form and matter. To understand the connection to the previous point, we should recall that the critical subject derives the impulse to criticize second nature from its ability to perceive the non-identity or contradiction between the categories of second nature and the objects on which they are imposed. Even as the second nature of capitalism increasingly engulfs first nature as well as subjective lifeworlds, our ability to perceive this non-identity shows that the engulfment is not total. A critique that takes its point of departure in the non-identity between form and matter is destructive of Cartesian dualism because such a dualism simply opposes two reified forms that are both social constructions, that of nature and that of society. Both are reproduced in capitalism and imposed on human as well as non-human matter and are therefore destabilized by the recalcitrance of that matter—a recalcitrance sensed in the anxiety and surprise described by Bloch and Hegel.

To summarize, the Cartesian dualism is an idealist imposition of form on matter. The methodological procedure of the critical theory of nature prevents this dualism from being absolutized. The form–matter distinction should thus not be confused with Cartesian dualism, but is rather precisely what we should hold on to in order to criticize and reject such dualism.

The Structure of the Book

The purpose of this book is to prepare and clarify a theoretical framework based on critical theory that can help us critically analyze what present-day capitalist society is doing to nature and thereby to itself. The reader will notice that the book's various arguments hook into each other, problems sometimes being raised in one chapter only to find their resolution in another. Here I want to summarize the main points of each chapter, thereby giving an overview of the book's structure.

In Chapter 2, "Marx's Three Materialisms," I suggest that it is useful to distinguish between three quite different readings of Marx's materialism: causal, practical, and critical materialism. Although elements of all three forms of materialism can be found among critical theorists, I argue that the idea of critical theory is best expressed in *critical materialism*. As I described earlier in this introductory chapter, such a materialism does not aspire to be a science so much as a critical tool—a critique of the system rather than a conceptual system itself. Capitalism is viewed as a totality since it forms an interconnected whole, but this totality is negative since the goal is to criticize it and abolish it. This critique is carried out by confronting capitalist forms with the matter on which they are imposed and demonstrating the contradictions generated by this imposition. Adopting this approach, however, leads to what I call the problem of the outside: how does critical materialism relate to what is outside the negative totality, which includes human as well as non-human nature? The critical procedure presupposes an investigation of this outside but seems unable to carry out such an investigation by itself. The problem of the outside is in turn expressed in two more specific problems which I refer to as the Lukács problem and the problem of utopia. Showing how a critical theory of nature can overcome these problems is an important task for the rest of book.

Chapter 3, "Natural History and the Primacy of the Object," shows how Adorno took decisive steps in developing two core theoretical notions in critical materialism: the *primacy of the object* and *natural history*. The idea of the primacy of the object helps us conceive of nature as more than a social construction but without reifying it by identifying it with a system of natural laws or reducing it to something ineffable. The idea of natural history enables us to trace dialectical shifts in the concept of nature in a way that counteracts reification. The more we try to purify the category of nature, the more we are bound to discover historical elements in it, and vice versa. These two ideas provide the ground for understanding Adorno's use of concepts such as first and second nature, the domination of nature, and so on. They also, I argue, provide fertile ground for understanding contemporary phenomena such as the increasing entwinement of nature and history, the naturalization of artificial environments, and the role of ideas of the wilderness in environmental activism.

However, a problem that remains unsolved in Adorno is how his critique of identity-thinking and negative totality relates to *capitalism*. Does he, by focusing on a totalizing critique of transhistorical processes, fail to engage properly with capitalism and does this contribute to negativism and political quietism, as some of his critics have argued? In Chapter 4, "Capitalism and the Domination of Nature," I argue that Schmidt contributes to clarifying the place of capitalism in the domination of nature. Schmidt's most famous work is *The Concept of Nature in Marx*, which originated as a 1960 dissertation written for Horkheimer and Adorno. This has been hailed as a seminal study but has also been subject to scathing criticism by eco-Marxists. I start by pointing to what I see as the important points in Schmidt's account—an erudite exposition of Marx, a clear eye for the relation between second nature and metabolism, and working out an analysis of capitalism suffused by the methodology of Frankfurt School critical theory—before reviewing the eco-Marxist criticism. Although much of the criticism falls flat, I concede the existence of a weak spot in his writings, centered on the question of how to imagine a reconciliation with nature in a post-capitalist society. Here we touch on the problem of utopia, which highlights the difficulty for a critical materialism, focused on the critical analysis of the present, to present utopian visions that might guide political action.

Delimiting the negative totality to capitalism raises the problem of how to understand the interface regulating capitalism's interaction with nature. Chapter 5, "Marx, Value, and Nature," describes this interface through the lens of the Marxian value law. I discuss the question of how to understand nature's contribution to the valorization process and emphasize that this is a non-valued contribution. Far from making the law anthropocentric, it helps us understand the distinctiveness of capitalism's domination of nature and how it interrelates with the exploitation of labor. The discussion highlights the complexities of nature's outside status: being an outside on which the system depends for crucial resources it is reproduced by the system precisely as an outside. Nature is therefore an outside in two senses: both as form and as matter. It is reproduced in the form of an outside but it is also an outside in the sense of matter that is never identical to the form imposed on it. The contradiction between capitalism and nature is not one between two internally unified entities, since nature is itself split and contradictory. Nature can therefore not be opposed to capitalism as a wholesome unity, as is sometimes done in the environmentalist camp. This, I argue, throws light on Marx's critical procedure and explains the function in his writings of concepts that may appear transhistorical rather than immanent to capitalism (such as nature, use-value, wealth, and metabolism). They bring out the non-identity between the categories central to the capitalist valorization process and the human and non-human world it is meant to valorize, but they do so without themselves being affirmed or essentialized.

Chapter 6, "Constellations and Natural Science," turns to Adorno's idea of *constellations* and suggests how it can be developed to provide a solution to the Lukács problem and more generally to the problem of the outside. Constellations are made up of concepts that encircle the object, illuminating it from various directions without being fixed in a logical relationship to each other. What keeps them together is their ability to illuminate the contradictory nature of the object. In *Negative Dialectics*, Adorno presents them as an alternative to the identity-thinking that subordinates the objects to the requirement of systemic consistency. This alternative is possible since the subject is never wholly engulfed in the system. Being exposed to the experience of non-identity, the subject needs constellations to think the contradictions that it senses between the system and its outside. Constellations enable us to do that without having to subsume this experience to the categories of the system. Importantly, constellations can include elements from natural science and thereby point to a solution to the Lukács problem.

Having sketched the basic arguments of the critical theory of nature in the previous chapters, I turn to discuss alternative theoretical traditions in the following three chapters: the eco-Marxism of John Bellamy Foster, the world-ecological approach of Jason Moore, the new materialism of Jane Bennett, and the object-oriented ontology of Timothy Morton. Chapter 7, "Eco-Marxism's Return to Marx," criticizes weaknesses in Foster's attempt to develop a dialectics of nature capable of serving as a unifying methodological platform spanning nature and society. Rather than showing how dialectics ensures methodological unity, he oscillates between a model of natural praxis applicable to how we relate to nature from a lifeworld perspective and a model of dialectics as a contemplative natural science. I also point to misunderstandings that underlie his criticism of the Frankfurt School and demonstrate that critical theory and eco-Marxism are based on differing conceptions of dialectics.

Chapter 8, "World-Ecology and the Persistence of Non-Cartesian Dualism," focuses on the contributions of Moore. His studies of capitalism's dependence on "cheap nature" usefully illuminate how processes of exploitation interact with processes of appropriation but fail to consistently carry out his own program of surmounting dualism. The result, I argue, is a picture that is most convincing where it betrays its own self-professed monism. Chapter 9, "New Materialism and Dark Ecology," turns to approaches outside the field of Marxism proper that, like Moore, deny or deconstruct the nature-society duality but, unlike him, do so without taking their point of departure in a theory of capital. Both Bennett and Morton laudably point to the agency, vitality, or queerness of the non-human world, but pay insufficient attention to the processes through which it can be *reified*, meaning that things are deprived of vitality by being reduced to fixed attributes relevant to capitalism. I conclude that the problems with these approaches stem from their premature attempt to deny the nature-society boundary.

The final chapter, "Utopia, the Apocalypse and Praxis," turns to the last remaining of the problems mentioned above, the problem of utopia, and argues that a critical theory of nature is eminently compatible with praxis as well as a utopian imagination. Negative dialectics shouldn't be associated with a disempowering vision of total integration. The possibility of experiencing non-identity is also the possibility of a critical subject. Praxis can be conceived of as an interplay between subject and object in which experiences of non-identity stimulate self-reflection. Negative dialectics therefore goes well with action as long as the latter does not rigidify into a system insensitive to the objects with which we interact. The prominence of utopian visions of reconciliation in the Frankfurt School can be explained by recalling that utopias are critical instruments rather than blueprints.

To round off this first presentation of the outlines of the critical theory of nature, let me comment briefly on the three terms in the book's subtitle: capital, dialectics, and ecology. As for the word *ecology*, I employ it as a critical, rather than scientific, term. It stands for the interrelatedness of nature including human beings, the wider material context on which capitalism is parasitically dependent. As for *capital*, it is the object of critique, the self-moving substance of the capitalist system that continually seeks to accumulate but which measures this accumulation only in terms of its own metric, that of value, while disregarding everything else.[13] *Dialectics*, finally, is the tool by which the critique of capital is carried out. As the reader will notice, I use the word "dialectics" in three ways, all of them derived from the Frankfurt School. First, it is a mode of presentation that shows how the meaning of individual concepts is mediated by other concepts. In this sense, for instance, we can say that Marx's presentation of the basic categories of *Capital* is dialectical. Secondly, as negative dialectics it is a critical movement of thought that disrupts conceptual systems by confronting concepts and objects. Finally, dialectical relations of a more benign kind can also exist between subject and object, as suggested by terms such as "mimesis" or "constellations" in which concepts counterbalance each other to bring out an object in a non-repressive way.

2

Marx's Three Materialisms

To apprehend a crisis as ecological we surely need to be materialists, at least in the minimal sense of recognizing that everything we think of as social or cultural depends on material factors. But what might materialism mean apart from that? We all know that Marx was a materialist, but there is a surprising amount of disagreement about how to interpret his materialism. Often it is said to consist in viewing the economy as a base on which a superstructure of ideas is erected, but we also know that many Marxists—not least in the tradition of critical theory—have rejected this metaphor without ceasing to call themselves materialists. In what sense, then, was Marx a materialist, and what might the answer mean for our attempts today to use his theories for an analysis of ecological crises?

It is customary to distinguish between more objectivistic varieties of Marxism (often regarded as orthodox) on the one hand and more subject-oriented or "humanistic" varieties (associated with Western Marxism) on the other. Whereas the former varieties have commonly been characterized by a *causal* materialism relying on the base-superstructure metaphor, the latter have relied on a *practical* materialism centered on human praxis. However, this rough distinction misses an important third materialism in Marx's works, namely, *critical materialism*. In this chapter I want to highlight the difference between these three strands of materialisms and in particular I want to bring out and elaborate on the characteristics of the third, which is central to the critical theory of nature.[1] As I will show, each strand throws a different light on Marx's discussions about nature, and each is linked to distinct conceptions of dialectics, strategies for criticizing capitalism, and attitudes to natural science.

Although all three strands can find support to some extent in Marx's writings, the point of my discussion is not to engage in the debate about the overall unity of his thought. Rather than breaks I prefer to see shifts in his thinking. While practical materialism is prominent in his early writings, mature works like *Grundrisse* and *Capital* are closer to critical materialism. Causal materialism finds its strongest expression in Engels but is to some extent present also in Marx. My purpose, however, is not to trace how Marx's thought develops over time, but to bring out a clear picture of the three strands and highlight their differences. Doing so will help clarify the theoretical point of departure of the critical theory of nature and make it easier to understand the debates and disagreements that have erupted between proponents of the three strands. At the same time, I do not wish to suggest that the strands are watertight compartments. While they differ by the relative emphasis they place on causal analysis, praxis, and critique, no strand excludes any of these three elements.

Each strand has strengths and weaknesses. Recurring problems in causal and practical materialism have revolved around determinism and agency in the case of the former, and the limits of a subject-centered dialectics in the case of the latter. A central problem in critical materialism concerns its relation to the world outside the system that is targeted for critique—an outside that includes nature. Exactly how is it possible for a critical materialism, which aims at an *immanent* critique of capital, to illuminate the relation between capital and nature? This and related problems set the stage for the following chapters where I hope to show how these problems can be overcome.

Causal Materialism and the Dialectics of Nature

Central to causal materialism is that it endows material factors priority as *causal* factors. The basic idea is expressed in Marx's famous metaphor of base and superstructure. The relations of production, he writes in preface to *A Contribution to the Critique of Political Economy* (1859), constitute an economic base "on which arises a legal and political superstructure" and that "conditions the general process of social, political and intellectual life" (Marx 2010c: 263).

When interpreting this statement, I think it is important to remember its programmatic character, and that it is hardly followed up by or put to work in much actual causal analysis in Marx's works. To make sense of it we should see it in the context of Marx and Engels's critique of idealism. Its point is polemical, namely, to reject the autonomous causal power of ideas: ideas do not shape or determine reality but the other way round.[2] This is also the point of the equally famous passage where Marx claims to invert the Hegelian dialectic by standing it on its feet. While to Hegel thinking is "the creator of the real world," for Marx "the ideal is nothing but the material world reflected in the mind of man, and translated into forms of thought" (Marx 1990: 102).

Taken by themselves these passages are not proof that Marx is best understood as a causal materialist or that causality was his main interest. For causal materialism in the strict sense to emerge, the idea of a material base reflected in the superstructure or in mind must be complemented by a second idea, namely, that science should take causal analysis as its *chief* area of inquiry. This happens in the so-called scientific Marxism that developed among Marx's followers, most prominently in the Second International and in Marxism-Leninism. A good example of such causal materialism is Rudolf Hilferding's description of Marxism as a value-neutral science, solely interested in the discovery of causal laws: "so far as Marxism is concerned the sole aim of any inquiry – even into matters of policy – is the discovery of causal relationships" (Hilferding 1981: 23).

The development of scientific Marxism represents a far-reaching shift in the meaning of dialectics compared to Hegel. To see this, we must take issue with the common tendency to interpret Marx's inversion of Hegel as a simple reversal of the direction of causality between ideas and matter. This interpretation is misleading since it takes for granted that causality is the main concern of dialectics. Hegel's main concern, however, is hardly with causal explanations.[3] To him, dialectics is better described as a retrospective recreation of meaning that unfolds through concepts than as a causal force operative in history. What he calls determination has less to do with causal, than *conceptual*, necessity. The relations between the moments that constitute the meaning of a totality (such as world history or the system of right) are "necessary" since they are "founded in the *inner identity* of the Notion" (Hegel 1969: 806). In other words, when Hegel speaks of necessity he is not invoking a relation of cause and effect so much as stating that certain conceptual relations are necessary in order to account for the meaning (*der Begriff*, or the Notion) of the totality under question. His idealism springs from his belief in the ability of thought to fully grasp these relations, which to him define reality, not from endowing ideas with causal primacy.

I stress the conceptual orientation of the Hegelian dialectics not to criticize those who make a causalist reading of Marx but to problematize the idea that the latter's materialism arises through a simple inversion of the direction of causality in Hegel. To the extent that Marx stresses causality in passages such as the one about base and superstructure, these passages *shift the meaning* of dialectics as such, from an emphasis on conceptual determination to an emphasis on causality.

With the shift from Hegel to scientific Marxism, conceptual relations in the realm of thought are thus translated into causal relations between things. The results of this transformation are visible in Engels's *Anti-Dühring* and *Dialectics of Nature*, which had a tremendous influence on subsequent Marxist thought.[4] Their main points are easily summarized. Overall, they offer a panorama of scientific socialism, in which the elements of what would later be known as orthodox Marxism are present: a cult of science, a historical teleology buttressed by a causal conception of dialectics, and a Promethean faith in the growth of the productive forces. Their dialectics is contemplative and objectivistic—it "is nothing more than the science of the general laws of motion and development of nature, human society and thought" (Engels 1987a: 131). Its three laws—those of the transformation of quantity into quality, the interpenetration of opposites, and the negation of the negation—are portrayed as operative in the world even without the intervention of human subjects.[5] Engels emphasizes that they apply to both nature and society, since both realms are characterized by change: "nature does not just *exist*, but *comes into being* and *passes away*" (Engels 1987a: 324).

The relation between Marx's and Engels's views on nature and dialectics is a subject of controversy.[6] While Marx was less contemplative and objectivistic than Engels's and more oriented to history and praxis, the fact that he approved of the *Anti-Dühring* shows that his position was ambiguous. In fairness to Engels, we should recognize that it is perfectly legitimate from a Hegelian standpoint to illustrate dialectics with examples from natural science. Hegel himself sprinkles the *Science of Logic* with passages that illustrate the dialectic with the behavior or chemical substances and planetary trajectories (e.g., Hegel 1969: 356ff, 369f, 379ff). The point where Engels breaks with Hegel is not in applying dialectics to natural science, but in portraying dialectics as a causal force operative in nature itself rather than as a retrospective reconstruction of meaning. The break, in other words, consists in his causalist or scientific reading of dialectics. While both Hegel and Engels exemplify what Lukács called contemplative thought, that is, thought disconnected from praxis, Marx differs from both through his ambition to materialistically *criticize* established forms. He doesn't just give matter pride of place, but he does so with a critical edge that tends to get lost in Engels's discussions of chemistry and physics.

Transforming dialectics into an instrument for causal analysis has implications for the defensibility of materialism. If dialectical relations are defined as causal, materialism is tenable only if convincing arguments can be presented for the causal primacy of material factors. Ironically, the insistence on a causalist reading of dialectics had the effect of making Marxism more rather than less vulnerable as a scientific endeavor since it came to be seen as requiring proof of the relative importance of various causal factors, a fact that in the course of the twentieth century led many Marxists to abandon the idea of a simple, one-sided, and mechanical causal relation between base and superstructure.[7] This softened or eliminated the determinism to which earlier forms of causal materialism had tended. The increasing readiness to recognize contingency and a complex interplay of causal forces reached an apogee in Althusser's structural Marxism. However, the fact that contingency must be carefully qualified if the materialist label is to be kept is illustrated by his repetition of Engels's formula that the material base determines history "in the last instance" (Althusser 1969: 112f, Engels 1890). The reason for this formula is plain to see: in a causal materialism, giving up the ultimate causal primacy of the base is tantamount to giving up materialism itself.[8]

Despite its vulnerability, causal materialism has had prominent followers not only in the east but also in the west. An example is Sebastiano Timpanaro, who defines materialism as "acknowledgement of the priority of nature over 'mind', or if you like, of the physical level over the biological level, and of the biological level over the socio-economic and cultural level" (1974: 7). This statement stands out for its explicit identification of the material base with nature and its acceptance of the objective validity of natural science. The targets of his criticism are those Western Marxists who out of a desire to avoid the taint of Stalinism, determinism, or "mechanical" materialism turned voluntarist and either abandoned materialism or identified the material base with the middle layer of social praxis instead of the real base of nature.[9] Somewhat similar arguments are presented by the contemporary eco-Marxist John Bellamy Foster, who praises the tradition of causal materialism for its respectful attitude to natural science and seeks to rescue it from the criticism it has received (see Chapter 7). While causal materialism has often gone hand in hand with a Promethean belief in technological progress and industrialization, the eco-Marxist version of causal materialism shows that such materialism can be given an ecological slant.

Regardless of the extent to which Marx adhered to a causal materialism, the aspiration among later Marxists to develop a scientific Marxism often led them to adopt a stronger causalist language than Marx himself—a transformation that gave rise to the popular image of Marx as a causal determinist.[10] This went hand in hand with a causalist reinterpretation of dialectics or, as in Althusser, to a rejection of dialectics as such. The reliance on science meant that this materialism tended toward a contemplative or traditional mode of theorizing that set up theory and praxis as external to each other. Dialectics turned into a tool for discovering laws of nature and society rather than for criticizing society. A dichotomy of science and ideology was erected, in which the delusions of the latter were denounced from the vantage point of the former. Not surprisingly, critics argued that causal materialism left little room for subjectivity, freedom, and political praxis.

Practical Materialism: Totality and History

Practical materialism is strongly present in Marx's early writings—the rejection of contemplative philosophizing in the "Theses on Feuerbach" is a clear example—but is perhaps best represented by the classical figures of so-called Western Marxism such as Lukács, Bloch, Korsch, Gramsci, Sartre, and others. The challenge that the practical materialists mounted against causal materialism took the form of a revival of Hegel. They argued that the base-superstructure model was undialectical since it neglected the subject–object relation and hence also praxis, in the sense of human interaction with the environment. However, since elements such as praxis and the subject were hard to apply outside the human realm, nature could only be grasped dialectically to the extent that it became involved in human action. Nature, from this point of view, is less an objective realm obeying its own dialectical laws than a counterpart of human labor that is constantly drawn into and reshaped by human practices.

In practical materialism, the material element is not an objectively given base, but human praxis as an integral part of a historical totality. The idea of the historical present as a totality through which all moments are mediated is central. Only by adopting the point of view of totality is it possible to grasp the mutual mediation of the acting subject and its environment. This is crucial to avoid reification, that is, the appearance that things exist independently of history and human praxis. Marx hints at the determining role of totality in formulations such as the following from the *German Ideology*: "It is not consciousness that determines life, but life that determines consciousness" (Marx & Engels 2010: 37). This formulation superficially resembles the base-superstructure metaphor, but instead of opposing each other as separate entities, the terms "life" and "consciousness" suggest a relation between totality and parts since consciousness is part of life rather than separate from it.[11] Implicit in the formulation is the idea of a whole, or a totality, in which consciousness is part along with praxis. Rather than assigning primacy to any part, the formulation suggests that all parts, including the economic activities constituting the so-called base, are mediated through a whole.

Wedded to the idea of totality is the elevation of dialectics into the distinguishing method of Marxism. Lukács famously declared that orthodox Marxism is nothing but the dialectical method, which consists in studying objects in relation to the totality in which they are mediated (1971: 1). This method alone was capable of breaking the reificatory grasp of "bourgeois" science with its rigid separation between form and content and its insistence that the essence of things can be grasped without regard for historical context. This dialectic is markedly different from the predetermined formulas laid down by Engels. Firstly, if everything is mediated through the totality there can be no neat separation of theory and praxis, and the idea of an objective, contemplative science must be rejected. Since praxis co-determines history, theoretical room is created for agency. "Men make their own history" but not "under circumstances chosen by themselves," as Marx put it (2010b: 103). It is the connection between praxis and totality that makes it possible for Lukács to praise Lenin as a thinker who because of his grasp of totality was able to act flexibly, with "revolutionary Realpolitik" (Lukács 2009: 18, 79–82). Precisely the perspective of totality allows for utmost flexibility in choosing the best means to reach the goal of revolution: "Without orientation towards totality there can be no historically true practice" (Lukács 2009: 96).

Secondly, this dialectic is opposed to the positivistic methods employed in natural science. Its aim is not to debunk delusions with the help of science, but to criticize reified thinking that treats things as if they had ahistorical essences. Unlike such thinking, dialectics shows that entities are historically constituted and changeable. The target of this criticism includes causal materialism, which is seen as reifying history itself by subordinating it to a set of deterministic laws. As Gramsci points out, Marx was never so cheap as to study dialectics only in order to search for ultimate causes (Gramsci 1971: 460).

Finally, we should note that although the dialectics employed by practical materialists is Hegelian in inspiration, it is less inspired by what Chris Arthur (2002) calls a systematic than by a historical dialectic. In other words, the aim of dialectics is to understand change rather than systemic interconnections. This is evident in Lukács, where it is mainly applied to understand the role of the revolutionary movement and the proletariat in the transition to socialism. Since the totality in relation to which this movement is mediated is not merely capitalism but history as such, it must also include the *negation* of capitalism, the class struggle waged by the proletariat.

Praxis and the Vision of a Unitary Science

Marx's so-called Paris manuscripts—a collection of notes taken in 1844—have been an important inspiration for practical materialists when it comes to the relation to nature.[12] A primary concern for Marx in these notes is how to overcome humanity's separation from nature in capitalism. Influenced by Engels, who in 1844 had written that socialism was "the reconciliation of mankind with nature and with itself," Marx describes communism as "the *genuine* resolution of the conflict between man and nature" (Marx 1981: 90). The notes provide a clear picture of his sociohistorical concept of nature, as an entity existing in conjunction with human praxis and labor rather than as a separate object in its own right. The concept of labor in these notes does not merely denote capitalist labor but is largely transhistorical. This comes forward vividly when in his dissertation he stresses the role of the senses—seeing, hearing, smelling, and tasting—as the medium through which the unity with nature is experienced: "In hearing nature hears itself, in smelling it smells itself, in seeing it sees itself" (Marx 2010a: 65). Rather than as an instrumental activity that reduces nature to an object of manipulation, labor is conceived as an active reshaping of nature through which humanity, which is part of nature, changes itself. The formation of the senses is therefore a product of history (Marx 1981: 96). It is thanks to our interaction with nature over the centuries that we can experience the beauty of certain sounds or flavor certain tastes. It is also thanks to this interaction with nature that humankind can realize its ability to consciously shape its world which distinguishes it from other species. This ability, however, is lost with the alienation of labor in capitalism, which disrupts the interaction with nature by depriving workers of control over production, the product itself, the social context of production as well as their own selves. The restoration of the unity with nature demands an overcoming of capitalism.

The Paris Manuscripts also envision a unitary science that would straddle the domains of history and nature. The vision shows that Marx's sociohistorical conception of nature has methodological implications. By viewing industry as a revelation of humankind's essential power

> natural science will lose its abstractly material – or rather, its idealistic – tendency, and will become the basis of *human* science, as it has already become – albeit in an estranged form – the basis of actual human life, and to assume *one* basis for life and a different basis for *science* is as a matter of course a lie ... History itself is a *real* part of *natural history* – of nature developing into man. Natural science will in time incorporate into itself the science of man, just as the science of man will incorporate into itself natural science: there will be *one science*.
>
> (Marx 1981: 98)

Unlike Engels, Marx opens the door for a dereifying critique of science in its present state. Since nature is the object of human activity and humanity part of nature, the separation of science into natural and historical branches is "abstract" and "idealistic." This idea influenced Adorno's idea of natural history, which I will discuss in the next chapter. Unlike him, however, Marx clearly believes that a scientific standpoint can be found from which to grasp nature and history concretely, in their dialectical interrelation. Exactly what this science is supposed to be is not spelled out, although it is clear that sense experience and sensuous need will play a key role in it (Marx 1981: 98; see also Feenberg 2014: 46f). We should note that Marx's vision has a utopian quality; he is not claiming that a unitary science is possible here and now. On the contrary, he suggests that such a science presupposes a standpoint in which alienation has been overcome (Mészáros 1970: 113f). As Alfred Schmidt puts it with blunt clarity, the unified science envisioned by Marx in the Paris Manuscripts will not come into being until communism. "Science" should not be confused with the positivistic "science" dominant today, or with the attempts to work out a unified science by way of "syntheses" (Schmidt 1973: 41f).

The passage shows that Marx was prepared to extend dialectics to the realm of nature. There are, however, crucial differences to how this idea was later realized by Engels in the *Dialectics of Nature*. Marx is not claiming that dialectics is a readymade instrument that can be applied as it is. Furthermore, to him nature is dialectical because of its integration with human praxis, not because of objective dialectical laws that would be operative even without human subjects. In his early manuscripts, Marx rejects the contemplative stance characteristic of causal materialism. By disregarding praxis, such materialism becomes abstract and idealistic. This is not to say that practical materialism is the sole correct lens through which to read Marx. In his later writings, he becomes more respectful of the natural sciences, not only refraining from criticizing them but often referring to them in support of his own arguments where relevant. "No natural laws can be done away with," he writes in a letter in 1868, "What can change in historically different circumstances is only the *form* in which these laws assert themselves" (Marx 1868). It is also the mature Marx who grimly dismisses his earlier hopes for communism as a definite solution to the antagonism between humankind and nature, and writes that nature will always remain a "realm of necessity" with which humans must "wrestle" in all societies to satisfy their wants (Marx 1991: 958f).

The Lukács Problem

Marx's vision of a unitary science set the stage for a recurring problem in Western Marxism during the twentieth century, namely, how dialectics should relate to natural science. The Western Marxist standpoint that dialectics is necessarily related to human praxis results in what has been referred to as the Lukács problem (since it comes forward in sharp relief in Lukács's writings; Foster et al. 2010: 224). Nature and natural science seem to become a domain where dialectics is powerless to criticize reification. If dialectics must halt before nature then that seems to imply that non-dialectical or positivistic methods are legitimate in natural science. This in turn seems to contradict Lukács's central argument that dialectics is the only way to grasp totality and break the hold of reificatory bourgeois science. In spite of his intentions, Lukács appears to land in a dualism according to which history and nature require different methods.[13] Steven Vogel (1996) claims that the entire tradition from Lukács to the Frankfurt School has been dogged by this problem. While Western Marxists had good reasons for rejecting Engels's objectivistic dialectics, in Vogel's view they failed to offer a better way to come to terms with nature, oscillating instead between restricting dialectics to society, defining nature as a social construct, and treating it as an ineffable other.

A fuller treatment of this problem will have to wait until Chapter 6. Suffice it to say here that a variety of responses have been proposed. Vogel himself has pushed for a thoroughgoing social constructivism (Vogel 1996). Other proposed solutions include defending Lukács's dualism of methods (Arato 1972: 38–43, Feenberg 1999, 2014, Jay 1984: 115f), extending the subject-category to nature (Bloch 1995: 672f), reconstructing a dialectics of nature in a less deterministic form (Foster 2000), and deconstructing the dichotomy between nature and society from the standpoint of assemblage theory (Loftus 2012: 62–6). As I will show in the book, objections can be raised against all these proposed solutions, but that doesn't mean that no solution exists. In contrast to Vogel, I will argue—in Chapter 6—that the Frankfurt School *does* provide a solution to the dilemma. This solution, however, becomes visible only if we shift perspective toward a critical materialism.

Critical Materialism and Negative Totality

In contrast to the revolutionary opportunities that Lukács saw when writing *History and Class Consciousness*, Frankfurt School critical theory was formed in a world characterized by the consolidation of oppressive systems. Rather than relying on a philosophy of history that preordained the proletariat to play the historical role of overthrowing capitalism, a more urgent task appeared to be to develop a critique that would not itself be reifying or likely to be recuperated into the ideological defenses of the status quo. Adorno played a pivotal role in these theoretical developments. Although his own efforts were mostly culturalist in orientation, younger researchers around him were inspired to take up the study of Marx, resulting from the late 1960s onward in the emergence of the so-called new Marx-reading. In this current of thought, which stretches from Adorno's negative dialectics to current Marx-scholarship, critical materialism comes fully into its own for the first time as a distinctive approach in the field of Marx-studies.[14]

Critical materialism rejects not only the base-superstructure metaphor but also the foundation of materialism in praxis. Rather than constructing a philosophical system where nature, matter, or praxis is given pride of place, it tries to reflexively surmount the tendency of thought, including materialist philosophy, to stabilize itself in the form of such a system. From the point of view of critical materialism, a materialist *system* is, strictly speaking, impossible since matter is not wholly subservient to thought. As Horkheimer points out, "even the assertion of the primacy of nature conceals within itself the assertion of the absolute sovereignty of spirit" (2013: 169). In critical materialism, matter is not a foundational concept, but exists outside conceptual systems, as a force capable of disrupting them. To paraphrase Fredric Jameson, matter needs no particular theoretical justification since it is "what hurts" (1981: 102). Hence critical materialism is less about making the concept of matter central than about sensitizing us to the ideological character of seemingly self-sufficient systems.[15]

Central to critical materialism is that matter makes itself *felt*, and that this can be utilized for critique. This doesn't mean that it simply relies on immediate sense-impressions. Impressions may provide an impulse or starting point for critique, but the critique itself must be carried out with concepts, through a so-called immanent critique. Immanent critique is often understood to mean that conceptual systems are criticized solely from within, but this is not entirely true. As Adorno points out, immanent critique involves a transcendent moment in the sense that it confronts concepts with the experience, the "objects," that they aspire to cover (1981: 17–34). It is therefore not entrapped within the criticized system but moves between it and the objects, gaining momentum from the clash or sensed non-identity between them.[16]

What makes critical materialism critical not just of idealism but also of capitalism is that the latter behaves like an idealist system that has become ingrained in society. What idealist philosophy does in thought—namely presuming its categories to successfully subsume reality—capitalism does though the market.[17] Adorno suggests that it is due to a "conceptuality" or idealist quality inherent in capitalism itself that capitalism can be reconstructed as a logic in the first place (1976: 79f; also see 1974: 22f). Far from being basic, the economy is itself seen as something spiritual that can be criticized by being confronted with its material outside—an idea that is central to Adorno and which we will return to in the next chapter. The task is thus not to criticize ideology as unreal compared with the genuine reality of the economic base, but to criticize the economy itself as realized idealism. In that endeavor, materialism is not a theory of matter but a critique of the dominant categories in the name of the outside reality that they claim to represent.

The model for this endeavor is the critical analysis of the constitutive moments of capitalism in Marx's mature writings. As Marx explains, *Capital* is not a work in economics but a "critique of political economy," that is, of economic categories: "It is at once an exposé and, by the same token, a critique of the system" (Marx 1858).[18] Unlike in practical materialism, which seeks to overcome reification by recourse to a totality in which all the splintered moments of bourgeois knowledge are reintegrated and given meaning in the light of the whole, critical materialism opts for an immanent critique that confronts the reified categories with what they purport to cover and thereby seeks to destabilize and disrupt them. Although these categories have become *de facto* predominant because of the historical emergence of capitalism, and in that sense "real," they should not be seen as ontologically valid for reality itself. Dialectics, as Adorno writes, is the ontology of the false state—that is, of a false reality that should be abolished (1975: 22).

Connected to the immanent approach is a further characteristic of critical materialism, namely, that it usually avoids applying dialectics in an abstract, transhistorical fashion.[19] Again, the model is Marx's mature works, where dialectics is primarily used to reveal the conceptual mediations that constitute capitalism rather than to explain historical transitions or make predictions. Dialectics thus plays itself out within a totality limited in time and space, namely, capitalism. As several commentators have pointed out, what corresponds to the Hegelian spirit or self-constituted subject in Marx is capital, rather than the proletariat, mind or humanity.[20] While this is true, it is important to stress, firstly, that this totality is not absolute, as in Hegel, but finite and dependent on external preconditions, such as nature and labor, which it is unable to secure fully by its own powers (Elbe 2018). Secondly, unlike in Hegel or Lukács, this totality is not a standpoint with which to identify, in order to grasp the meaning, or truth, of history. Instead, it is *negative*, meaning that it is deceptive and should be criticized and ultimately abolished.[21] As Adorno puts it, "the whole is the false" (Adorno 1978: 50). It presents itself as harmonious and meaningful, but only by covering up its own contradictions—for instance, when markets are presented as capable of generating value independently of the use-values supplied by labor and nature.[22]

The Problem of the Outside

Critical materialism is no less free of problems than other varieties of materialism. Above all, delimiting totality to a negative one raises the methodological problem of how to think the relation between this totality and its outside. Since this outside includes nature—for instance all the flows of nutrition, energy, and other resources required to keep the industrial machine running—the question is obviously crucial to a critical theory of nature. The problem, however, is that the methodological approach of critical materialism is so oriented to a critique of the totality that it appears less well equipped to thinking the outside itself. Isn't there a risk that a restrictive focus on the economic forms that make up the totality leads to a neglect of nature? Can a critical materialism even theorize nature, given that nature is external to the negative totality on which it focuses attention and aspires to criticize immanently? This problem has ramifications for the range of applicability of dialectics. Critical materialism uses a dialectical method of presentation to explain the constitutive relations *within* the totality, but posits the central contradiction *between* the totality and its outside. Can the relation to the outside then be thought dialectically or is dialectics strictly limited to the exposition of the negative totality? If the latter, doesn't this leave the relation to nature untheorized?

The problem of the outside is not just a theoretical problem but relates to the fact that it is empirically hard to specify a clear boundary to capitalism. Whereas Marx to a great extent succeeded in providing a clear model of the logic of capital—an ideal type of capitalism, so to speak—we hardly ever experience capital in this pure form in reality. What, for instance, are we to think of those domains of modern society that are actively *defined* as an outside by capitalism—such as nature, the family, or the state—on which capitalism remains dependent for the reproduction of labor power, the maintenance of social norms, and natural resources? Are such domains—which could be referred to as the backstages of capitalism, hidden from view but nevertheless contributing to its functioning—part of the system or not?[23] As a matter of experience, we always encounter capitalism embedded in or accompanied by a range of institutional arrangements that support it but which may shift over time or depending on place. This includes patriarchy; forms of state power such as the judiciary, the police, and the military; as well as ideologies that legitimate capitalism and delegitimate dissent. If, like Adorno, we attempt to find the impulse for criticizing the system in the experience of the non-identity between its categories and the objects to which they are applied, then it obviously matters that we hardly ever encounter the system in pure form. The categories that generate the sense of non-identity are not only the economic ones laid bare by Marx in *Capital*, but can also be categories central to the operation of the state, the family, or the various constructions of national and racial belonging that pervade our societies. For a value-form analysis to be useful to comprehend the reality of capitalist societies, the question of how to relate these backstages to capitalism is clearly central.

In my view, the Marxist tradition offers four basic options for how to theorize the relation between the capitalist totality mapped in *Capital* and its outside. Below I will indicate why I believe all of them are useful and practicable, to a greater or lesser extent. The first three, however, present difficulties that can be solved only when they are combined with the fourth option.

The first option makes use of what might be called a Hegelian retrospective positing of presuppositions. It takes its point of departure in the workings of capital as analyzed by Marx and carefully attempts to draw conclusions from it about capital's external preconditions. Marx himself made use of this option, as evinced by his argument that it is possible to gain insights into the vanished social formations out of the present one, just as it is possible to draw conclusions about the "anatomy of the ape" from the "anatomy of man" (Marx 1993: 105). Nancy Fraser's theoretization of nature, the state, and systems of reproduction as functionally necessary background conditions of capitalism is a contemporary example (Fraser 2014, Fraser & Jaeggi 2018). Common to Marx and Fraser is that they start from a critical analysis of political economy and draw conclusions about the environment outside the economic forms. Conducting such an analysis therefore doesn't mean that the range of dialectical analysis must be limited to capital itself. Instead, they are arguing that the logic of the system presupposes a certain relation to the reality outside the economic forms. Another clear example of this way of reasoning is when Werner Bonefeld argues that so-called primitive accumulation is part of the "conceptuality of capital." Rather than being a historical cause of the genesis of capitalism, it is constantly reproduced as a permanent feature of capitalism (Bonefeld 2014: 81).[24]

While this logic of positing the system's presuppositions is useful as far as it takes us, it doesn't solve the problem of how to theorize the relation to the outside. Basically, it amounts to pushing the boundaries of the totality outward, by making a certain mode of relating to the non-economic outside integral to it. This results in a widening of the capitalist totality, which in addition to the economic forms proper also comes to include a more nebulous outer layer of seemingly non-economic phenomena. However, pushing the boundaries does not solve the problem of how to think what exists outside these boundaries. The drawback of this Hegelian strategy of positing presuppositions is that it works best when mapping those relations to the outside that are functional for or constitutive of the system. But are relations to the outside only determined by the requirements of the system? Doesn't this strategy miss the primacy of the object, the fact that—as Adorno put it—reality never wholly goes up in the concepts? The logic of positing presuppositions helps us see why nature is useful to capitalism, but conflicts and other processes that appear non-functional to capitalism—nuclear meltdowns, pandemics, and resource depletion, for instance—have to be explained by other means.

The second option is that of theoretical eclecticism. The limited usefulness of the Hegelian strategy means that it is hard in actual research to avoid bringing in additional theories, such as those of natural science, to supplement the analysis of capitalism.[25] This move is reasonable but not without its problems. We easily end up in variants of the so-called Lukács problem concerning how a dialectical Marxist approach can be reconciled with non-dialectical theories. Even if bringing in such theories as a supplement is not bad per se, the question is *how* the Marxist dialectic should relate to them. If theories of nature are simply added to the dialectic, then we are well on the way toward what Horkheimer called traditional theory. If theories of nature are to be brought in, then this should be done in a way that preserves the critical incisiveness—but how this is to be achieved has hardly been clarified.

A third option is that of shifting levels of abstraction. Bertel Ollman is well known for the argument that dialectics in Marx always involves a multiplicity of levels of abstraction and that limiting oneself to the single level of the historically limited totality constituted by capitalism is wrong. According to this argument, nothing prevents us from raising the level of abstraction to a transhistorical one more suitable for investigating other historical modes of production or humankind's universal metabolism with nature (Ollman 2003: 182–92). As with the previous two options, the idea of shifting levels of abstraction is helpful to an extent. Marx himself makes use of such shifting when he juxtaposes concepts internal to capitalism (such as value or abstract labor) to concepts with a transhistorical application (such as use-value or concrete labor). The usefulness of this option, however, is limited by the weight of real abstractions. That Marx makes *statements* on various levels of generality is true, but it is only on *one* of these—the level of capitalism—that they come together in a totality, that is, can be grasped as constitutive moments of a notion.[26] Unlike transhistorical abstractions, those that bring capitalism into view are not simply intellectual tools but are anchored in social practices and institutions and hence impose themselves as real. Since capitalism is a system not only in thought but also in reality, it is misleading to suggest that we are free to choose the level of abstraction as we please.

Having run through the first three options, we appear to have arrived at a trilemma. The first option of using the Hegelian logic of positing presuppositions remains incomplete as a strategy for grasping the system's outside in its own right. The second option of eclecticism leaves the problem of how to interrelate the various parts of the theoretical bricolage unsolved. The third option of varying levels of abstraction can be misleading in suggesting that abstraction is nothing but a mental operation.

A fourth option, however, becomes visible when we turn to Adorno's negative dialectics. Here the point of departure is the experience of non-identity between concept and object, or in other words in the pain generated when the negative totality of capitalism imposes itself on the material world. This is a world in which we are already part, by virtue of being alive, and to which there is therefore no need to reach out. Here we cannot ignore the significance of real abstractions, for it is as real—in the sense of being embodied in practices and institutions that we encounter in our lives—that the categories of capitalism generate pain. From that pain a critical impulse is generated that is indispensable in immanent critique. In Chapter 6, I will argue that Adorno's idea of constellations helps us see how this impulse opens up the possibility of a conceptual thinking about nature. There I will also discuss how this fourth option can be combined with the other three options.

Nature and the Three Materialisms

In this chapter I have distinguished between three ways of interpreting Marx's materialism—as causal, practical, and critical (Table 1). Each of them has strengths as well as weaknesses, and each casts a different light on the problem of nature.

Table 1 Three interpretations of Marx's materialism

	Causal	**Practical**	**Critical**
Pertinent cross-reference to Marx	The "Preface" (1859) on base and superstructure	Consciousness determined by "social being" or "life"	*Capital* as a "critique of political economy"
Developed by	Engels, Bukharin, Althusser, Timpanaro, and so on.	Lukács, Gramsci, Sartre, and so on.	Adorno, Schmidt, Postone, the new Marx-reading, etc.
Core ideas	• The base is reflected in the superstructure ("reflection-theory"). • Causal determinism, which is ultimately ("in the last instance") seen as one-way.	• All moments are mediated through the totality. This totality can only be grasped dialectically and includes our own praxis. • Rejection of reified "bourgeois" thought. The attempt to grasp moments in isolation from the totality is tantamount to reification.	• The logic of capital forms a negative totality, locked in a relation of non-identity to its outside. • Rejection of the pretense of thought to capture reality—a pretense that is idealist even when dressed up as materialism.
Mode of determination	Causal determinism (which can be supplemented by a recognition of contingency and complexity)	• Praxis co-determines history. • The dialectical method is applied transhistorically and includes class struggle.	Dialectics refrains from transhistorical accounts. Only the system's internal, constitutive relations are determinate.
Type of dialectic and relation to Hegel	• Dialectics as causal interaction. • Hegel is "inverted" or rejected in the name of science.	• Historical dialectic in which the "spirit" is embodied in the proletariat. • Entities are not fixed, but historically constituted. Dialectics is a tool for understanding historical change.	• Systematic dialectic in which totality is negative. The "spirit" is embodied in capital. • Hegel's and Marx's main works aimed at explicating the logic of given wholes.

Mode of critique	• Dogmatic • Science vs. ideology	• Dialectics vs. reified thought • The "immediate" is shown to be mediated through totality.	• The non-identical vs. identity-thinking • Negative dialectics: The conceptual whole is undermined by its relation to its object.
Theoretical stance	Contemplative / traditional theory	Unity of theory and praxis	Immanent critique, critique is internal to the presentation.
Weaknesses	The role of agency	The Lukács problem	Relation to the outside

If Marx is read as a causal materialist, his remarks on nature come forward as a preliminary scientific inquiry into the natural factors or conditions on which human society depends. Nature is seen as the most fundamental level within the overall base-superstructure model, even more basic than that of productive forces or relations of production. The weakness of this approach is that Marxism tries to behave as a quasi-natural science, competing with established natural science. If the validity of the natural sciences is taken for granted, as in Engels, it becomes hard to see what dialectics can add to them. In addition, this approach often yields a rigid, reified view of both nature and history by adopting a contemplative stance to its subject matter that seems to leave no room for agency.

If Marx is instead read as a practical materialist, utopian possibilities open up to the extent that nature is recognized as a living, congenial environment deserving of friendship, or even as a subject in its own right, as in Bloch. The weakness of this approach is that nature may disappear behind the human meanings attached to it, thereby eradicating non-identity. In addition, it wrestles with the methodological dilemma of how to extend a dialectical method that is tied to the human subject and to human praxis to a subjectless natural realm.

In contrast to the above readings, critical materialism stresses the *critical* intent behind Marx's references to nature. The relation to nature is mobilized as a critique of capitalism's semblance of self-sufficiency, rather than to create a science of how nature affects society or probing into its meaning for acting subjects. To be a critical materialist is to criticize capitalism by confronting it with its excluded other, with the non-identity between its concepts and reality.

Even when the system presents itself as a seamless and self-sufficient totality, we know it as oppressive and exploitative thanks to such experiences that help us see the non-identity between the dominant conceptual categories of the system and the reality they subsume.

The following chapters will further develop my argument that critical materialism is the most promising way forward for a critical theory of nature. To do this they must show how critical materialism can solve what I have called the problem of the outside, that is, of how to illuminate the relation between capital and nature from the point of view of an immanent critique of capitalism. As I will show in the course of the book, this problem is connected to two further problems, one of which is the previously mentioned Lukács problem. How can critical materialism contribute to a solution to the question of how to combine dialectics and natural science? Must dialectics pull back from natural science and limit itself to investigating nature as it reveals itself to us in the relations of labor and aesthetic contemplation? A second problem concerns utopia and political action. How is it possible, from the standpoint of a critical materialism, to imagine a reconciliation with nature? While such a vision can be presented as an integral part of the philosophy of history in practical materialism, how can it be legitimate in an enterprise focusing on a critique of the present? A similar problem exists in relation to praxis and political action. The very act of taking a critical stance toward a totality seems to presuppose a subject that is not wholly subsumed within that totality. What, then, is the place of the subject in relation to totality, to what extent is it itself conditioned by it, and what are the possibilities of resistance?

These problems—the general one of the outside as well as the more specific ones related to natural science, utopia, and political action—are all relevant to how we should understand and respond to environmental destruction. In the following chapters I will argue that resources for solving them are provided by Adorno's negative dialectics and Schmidt's theoretization of capitalism.

3

Natural History and the Primacy of the Object

Adorno had a well-attested fondness for animals. He jokingly likened himself to a pachyderm, decorated his desk with animal figurines and a teddy bear, and wrote in a letter to Horkheimer that he wanted to lay the theoretical groundwork of a society that included animals (Mendieta 2011: 150). While this may have been little more than personal quirks, it is hard not to associate it with a core notion of his philosophy, that of the primacy of the object (*Vorrang des Objekts*). Other lifeforms are often his preferred examples when he argues the need to be true to an object and do it justice even where it goes against the dictates of rationality. "Philosophy," he states, "exists in order to redeem what you see in the look of an animal" (Horkheimer & Adorno 2010: 51). Withered trees, vivisection, circus animals, and the dwindling elephant herds of Africa are also mentioned as examples of suffering nature when he criticizes the cruelty inherent in instrumental reason.[1]

In this chapter I want to lay out two ideas, based on Adorno's works, that belong to the distinctive kernel of the critical theory of nature: that of the primacy of the object and that of natural history (*Naturgeschichte*). Together they refer to nature both as a material reality and as a historically shifting idea. Thereby they illuminate an ambiguity in nature, which is evident in our everyday use of the concept. On the one hand, nature exists beyond the world of our thoughts, and independently of it, and will never be fully known despite the progress of science. It is withdrawn from the subject—the mute object par excellence. But paradoxically, it also possesses an affluence of meaning, being one of the best examples of an almost purely social category that has been loaded with a variety of different meanings depending on historical and cultural context. With the help of Adorno, as I will show, we can see how these two sides of the concept of nature belong together.[2]

Below, I start with a general remark on my interpretation of Adorno before turning to the presentation of his idea of the primacy of the object. This idea expresses a core premise of critical materialism, namely, that the subject and its conceptual constructs are always secondary in relation to the objects that they attempt to master and control.[3] As Adorno phrases it in *Negative Dialectics*, the primacy of the object means that the relation between subject and object is asymmetric in the sense that the subject is not independent of the object but depends on it. The subject is itself always object, but the converse is not necessarily true. Just as nature can subsist without humans but not humans without nature, so the object can persist without thought but not thought without object (Adorno 1973a: 183f, 2005: 249; see also Cook 2011: 40f, Wiggershaus 1994: 602f). To Adorno, the object's primacy is not simply a postulate, but something that we ascertain experientially when we sense that our concepts do not fully grasp their objects, or, as he puts it, that "objects do not go into their concepts without leaving a remainder" (Adorno 1973a: 5). Such experiences of non-identity between concepts and objects are touchstones for his critique of identity-thinking, a mode of thought pervading philosophical systems as well as capitalist society.

In the chapter's second part, I turn to the theme of nature in Adorno's writings, beginning with the idea of natural history that is crucial to understanding his usage of the concept of nature. Nature is used in a variety of ways in his works. The reason for this is that nature, like other concepts, gains its meaning in relation to its ostensible opposite, history. He summarizes the dialectical interrelation between nature and history in a succinct sentence that has the form of a *chiasmus* (a stylistic device that consists of a reversal of terms in otherwise parallel phrases): just as history solidifies into nature, nature reveals itself as historical. Here I use the idea that history and nature dialectically revert into each other to defend the *Dialectic of Enlightenment*, which Adorno coauthored with Horkheimer and which has become a classical reference point for the Frankfurt School view of nature. Against the popular understanding, I argue that this work shouldn't be read as a pessimistic philosophy of history. Its point is to undermine the fixed meanings of myth and enlightenment and show how they revert into each other in analogy with the concepts of nature and history. Its aim is not to proclaim the inevitability of regression, but to achieve a critical effect, thereby disrupting the processes it is describing.

Together, the idea of natural history and that of the primacy of the object enable Adorno to work out a concept of nature that is neither naively realist nor fully constructivist. The idea of natural history means that every mention of nature in his writings is ambiguous, referring to a moment in a larger dialectical movement. He rejects the realist belief that concepts correspond to a fixed, knowable reality. But his seeming constructivism only pertains to the world of concepts. Central to the idea of the primacy of the object is the materialist conviction that concepts are not everything. By showing how natural history and the primacy of the object work together, Adorno helps us conceive of a way to defend the materiality of environmental crises against constructivist critiques of the notion of nature.[4]

The final part of the chapter is devoted to the problem of how Adorno's account of nature relates to capitalism. Are critics right that he misses the crucial role played by capitalism in the destruction of nature, which instead tends to get explained by reference to a transhistorical process of an increasing domination of nature through instrumental reason? A variant of this criticism targets his attempt to take the critique of conceptual systems as a model for the critique of capitalism. Is it reasonable to posit an analogy between the functioning of conceptual and institutional systems? My argument will be that Adorno can be defended against these two criticisms only by means of considerable theoretical work to bring capitalism sharper into focus, much of which was left to younger critical theorists, such as Alfred Schmidt.

The Negative and the Positive Sides of Adorno

First a word on how I choose to interpret Adorno. Two quite different interpretations are possible, depending on how one views the relation between three prominent elements in his thinking. The first element is the idea that today's societies are subjected to an objective context of delusion (*Verblendungszusammenhang*), that is, that societies have become so totally integrated that external standpoints from which to judge, criticize, or resist the system are hardly discernible. The second element is his idea of negative dialectics, which is usually understood as an attempt to show how philosophy can function critically in the context of delusion, as an immanent criticism. The third element is the utopian idea of reconciliation, which he opposes to the prevailing society.

How should the relation between these three elements be interpreted? The problem is the tension between the first and third elements, between the thesis of the context of delusion and the utopia of reconciliation. The utopia represents a case where his philosophy seems to abandon a purely negative, immanent position and become affirmative. Although descriptions of this utopia are sparse, it indicates that he doesn't see the context of delusion as total. A context of total delusion would seem to rule out the possibility of even mildly affirmative visions of utopia since the latter would then have to be delusive as well.

In the face of this tension, a possible interpretation is to stress the link between the thesis of the context of delusion and negative dialectics while relegating the idea of reconciliation to the sidelines. In this interpretation it is due to the context of delusion that philosophy must be resolutely negative in order to avoid being used for ideological purposes. Here negative dialectic comes forward as simply a negative or non-affirmative procedure.[5]

This interpretation, however, fails to make sense of the idea of reconciliation, which becomes an anomaly. I therefore prefer a different interpretation that instead stresses the link between the utopia of reconciliation and negative dialectics. These two elements are not incompatible but presuppose one another, being two aspects of the same idea, namely, reflection on experience. Negative dialectics is a movement of thought between concept and object that takes its point of departure in the experience of non-identity between the two. While this movement appears negative from the point of view of the conceptual systems in which it triggers a breakdown, it also liberates the concepts by setting them afloat in a reflective movement in which they are guided by the primacy of the object rather than by the conceptual logic of the system.

An illustration of this movement of thought can be found in a well-known passage in the *Dialectics of Enlightenment* where Horkheimer and Adorno describe the mimetic relation to nature as an anticipation of a reconciled relation between subject and object. The state of true mimesis, they argue, has its source in a reflexivity that neither sticks to the immediate sense impression nor entangles itself in a repetitive systemic thinking, but moves between the two:

> Only in that mediation by which the meaningless sensation brings a thought to the full productivity of which it is capable, while on the other hand the thought abandons itself without reservation to the predominant impression, is that pathological loneliness which characterizes the whole of nature overcome.
>
> (Adorno & Horkheimer 1997: 189)

Here the reflection on experience is presented as the remedy for the blindness to context that results both from identity-thinking and self-surrender to the immediate impression. Its role is not just destructive, but also suggestive of a possible reconciliation. The negative, immanent movement of critique reveals itself, surprisingly perhaps, as utopian, as a model for interacting with the world that is open or porous regarding experience and guided by what Adorno (1978: 247) calls the "felt contact with its objects." From this perspective, negative dialectics is the attempt to be true to utopia by doing justice to the experience of non-identity.

When I emphasize the importance of experience as a touchstone for critique in this book, I include the movement of thought between subject and object described in the passage above as a central connotation of experience. Experience should not be confused with mere sense impressions, but arises where sense impressions and thinking are conjoined. Concepts are needed to contextualize the impressions, yet conceptual thought operating in isolation from the senses is insufficient for experience since it shuts itself off from the uniqueness of the objects, which can never be wholly captured by concepts.

The paradox that negative dialectics can have a positive or generative aspect disappears when we realize that it is only from the two criticized standpoints of the immediacy of sense impressions and of conceptual systems that negative dialectics needs to appear as a purely destructive procedure. Seen as a reflection on experience, it anticipates a reconciliation that would not be a state of unity or identity but rather what Adorno calls a "reconciliation of differences" or a state "in which people could be different without fear" (Adorno 1978: 103; see also 2005: 247). What falls outside the pattern in this interpretation is not the utopia of reconciliation, but rather the thesis of the context of delusion to the extent that the latter is thought of as total. I therefore want to argue that to Adorno capitalist society is in fact *not* a wholly integrated totality, although it is driven by a totalizing logic. Society may be *on the verge* of becoming an "open-air prison," but this is still not an accomplished fact (Adorno 1975: 34).[6] The very fact that experiences of non-identity are possible proves that all contradictions haven't been pacified. Negative dialectics takes its point of departure in these contradictions and is not a reaction to total closure. In this interpretation, Adorno is utopian rather than merely negative—a thinker in whom a slight hope for reconciliation is preserved even in the face of almost vanishingly slim odds.

The Primacy of the Object: Negative Dialectics as Critical Materialism

What does Adorno mean by object? Although he refrains from explicit definitions, he clearly thinks of them as *objects of experience*. They are not mysterious entities beyond the grasp of thought. A childhood memory can be an object, as can the subject matter of an artwork, a pollution incident, or the gaze of a suffering animal. Objects, then, are not necessarily physical things. Nor are they mere sense impressions or facts in a positivist sense. Their meaning is mediated by concepts. Like in Hegel, we can grasp the meaning of things only by the conceptual mediations that make up what we feel to be their essence. Phenomena like climate change, mercury poisoning, or a virus, for instance, are mediated through conceptual chains that include not only biology and chemistry but also court proceedings, economic charts, the experiences of victims, and much more. In other words, the meaning of objects is determined by concepts that dialectically determine each other's meaning.

At the same time, objects are *contradictory* since the concepts refer to experiences that can destabilize them.[7] Every encounter with a victim of mercury poisoning can change my understanding of it. Such encounters are determinate negations that impel me to add new concepts to my understanding to rectify the shortcomings of the ones previously employed. For example, when I consider my concept of mercury poisoning, I am likely to discover that a mere list of the symptoms included in medical definitions is insufficient. Due to the harrowing history of Minamata disease in postwar Japan—the single worst incident of mercury poisoning in the world so far[8]—my understanding of such poisoning is also colored by things like dancing cats and dead fish, decade-long struggles in court, the ostracization of impoverished fishermen from their communities, the disruption of shareholder meetings by desperate protesters impersonating ghosts, and mourning survivors carving Buddha statues out of stone, among other things. More than medical definitions, these historical experiences have become indelible parts of my concept of mercury poisoning—parts that I must seek to do justice when I describe the phenomenon. As a result, however, of letting the concept be colored by experiences in this fashion it may lose coherence and become self-contradictory. Hegel describes dialectics as a process through which we move from abstract and simple characterizations toward concrete and complex ones. Adorno would add that to grasp a thing concretely is to understand it through the unresolved contradictions in which it is entangled and which prevent its meaning from stabilizing into a unity.

Highlighting these contradictions is an essential part of Adorno's critique since they make us aware of power and oppression. The categories through which dominant institutions such as courts or corporations handle mercury poisoning are often glaringly at odds with how it is experienced by the less powerful. Often it is precisely from that discrepancy that the sense of non-identity arises that becomes the starting point for critique.

That concepts can be destabilized by experience means that they can never rest in final definitions. As mentioned, even a simple act like talking to a victim—or becoming one myself—can destabilize my concept of mercury poisoning. Such destabilizations, however, do not land me in relativism that distances me from the object and paralyzes my ability to act. On the contrary, they are driven by a fidelity to the object that opens up for an empathy with suffering.

Anti-idealism

The idea of the object's primacy is central to Adorno's criticism of idealism, understood as the pretense to fully capture reality through conceptual systems.[9] That concept and object are non-identical means that the world can never be fully grasped in a system of thought. This point also undermines certain forms of materialism, since it means that a viable materialism cannot take the form of a conceptual system. Such a system would still be idealist in form, even if it elevates matter to its first principle.

"It is by passing to the object's preponderance that dialectics is rendered materialistic," Adorno writes (1973a: 192). His *Negative Dialectics* can be read as a sustained attempt to work out a materialism that does not succumb to idealism. This is not done by abstractly rejecting systems, but by confronting them with non-identity, with the gap that separates object and concept. Prioritizing the object means resisting the tendency of concepts to subsume the objects. Instead, negative dialectics uses the encounter with the object to trigger a "logic of disintegration" of conceptual systems (1973a: 144f). Concepts that are free from the constraints of the system can instead arrange themselves in what Adorno calls constellations, a plurality of concepts that encircle the object and illuminate it from different sides without subsuming it.[10]

It should be clear that the primacy of the object doesn't mean objectivism in the sense of a naive realism that presumes its concepts to be adequate to reality and therefore is oblivious of non-identity. Nor does it mean eliminating subjective determinations in favor of a direct, unmediated perception of the object. As Adorno never tires of pointing out, such an appearance of immediacy obscures the processes whereby it is mediated, or given meaning, by its social context.

But how can objects be sensed, if concepts cannot fully grasp them? The answer is that experience isn't just conceptual but also sensuous. Just as in Marx, a somatic element is strongly present in Adorno's materialism. Adorno describes it as a "felt contact" with the objects (1973a, 1978: 247), which is especially keen in moments of suffering. His critical theory, it has been suggested, can be described as "sensitive theory" based, not on concepts or first principles, but on a standpoint of "*a priori* pain" (Sloterdijk 1987: xxxiii). It is from sensitivity to pain, not from any preestablished standpoint, that his criticism gains its incisiveness. This description is fitting since it depends on a highly developed, almost exaggerated, sensitivity toward objects to operate. Here we see the main reason for his persistent focus on suffering and kindred emotions such as shock, nausea, and vertigo (Adorno 1974: 83, 1975: 43, 202). They help us apprehend the non-identity between concepts and objects.[11]

The primacy of the object has epistemological as well as aesthetic and moral implications. Epistemologically, experience of the object provides a way to see through the lie of identity and break through the closure of the system (Adorno 1974: 178f, 1975: 43, 202). In aesthetics, the ability of art to salvage the experience of non-identity becomes a hallmark of its truth—only what does not fit in is true, as he puts in in *Aesthetic Theory* (Adorno 1997: 59). But the experiences also contain a moral imperative. "The possibility of pogroms," he writes in *Minima Moralia*, "is decided in the moment when the gaze of fatally-wounded animal falls on a human being" (1978: 105). When such a gaze falls on an onlooker, confronting the latter with its agony, a mimetic impulse can arise that changes the relation between subject and object. The animal is no longer a mere object but a subject that calls out to me for help. To the extent, however, that the modern human subject has been forged out of the repression of mimetic impulses, it is only capable of viewing its environment instrumentally, as mere objects to be manipulated. In contrast, the subject that empathizes with suffering recognizes itself as part of nature. The sympathy for the suffering animal brings about a "remembrance of nature within the subject" and, along with that, a remembrance of the suffering caused by our self-repression (Horkheimer & Adorno 2002: 32; see also Flodin 2011, Sanbonmatsu 2011b: 6f). The freedom that we may experience in such moments of empathy is an expression of the fact that our striving to dominate nature has gone hand in hand with a repression of our own inner nature, and that the empathy therefore paradoxically involves a momentary relief from the self-estrangement of our own animality.

The Rejection of Causal and Practical Materialism

The primacy of the object highlights the contrast between Adorno's critical materialism and other brands of materialism. The latter remain idealist to the extent that they rely on conceptual systems to capture reality. The problem with the base-superstructure metaphor, for instance, is that it is not materialist enough. Even if such a system is labeled materialist, it retains the form of an idealist system through its claim to have captured reality in a net of reified concepts. This claim presupposes the principle of constitutive subjectivity, the idea that thinking has "supremacy over otherness" (Adorno 1975: 201).[12] In contrast to such materialism, Adorno's materialism refuses to grant matter or nature a *place* in a system of thought as a base, essence, or first principle. Materialism, he writes, "cannot be a system" since matter does not follow from principles (Adorno 1974: 264).[13]

The aspiration to criticize idealist systems, Adorno argues, was also the core of Marx's own materialism, which didn't take the form of a *Weltanschauung*, an ontology or a theory of nature, as materialism did in the Soviet bloc. Marx, he argues, was a materialist since he criticized system-thinking, not because he established a materialist system. Rather than aiming for a new substantial economic theory, the aim of *Capital* was to criticize political economy. It was "a negative system," a critique of the systematic unity and consistency of capitalist society (Adorno 1974: 216, see also 255f).

The primacy of the object is also crucial to Adorno's criticism of the praxis-oriented materialism in Lukács's *History and Class Consciousness*. Adorno and Lukács both reject the base-superstructure metaphor, but from different standpoints. In Lukács's work the spell of reification was to be broken by the proletariat's revolutionary praxis, which alone opened up the possibility of grasping totality and restoring to humanity the ability to shape its world. This position was plausible in a time animated by the revolutionary fervor in Hungary and Russia starting in 1917, but to Adorno, writing in a subsequent period when revolutionary praxis seemed blocked, Lukács's work appeared entrapped in idealism because of its tendency to present all objectivity as flowing from the subject-as-producer, denying the object independent autonomy.[14] In that tendency, the principle of constitutive subjectivity reappeared. Fighting reification, Adorno insists, mustn't mean subordinating nature to the subject, but should involve cultivating a sensitivity to the damage done by the subject to the world outside it. This does not mean abandoning praxis. He is no less insistent than Lukács on the need to reawaken possibilities of historical change. However, he also stresses the need to attend to the opposite process whereby such change reverts into a new nature that reproduces the ills that need to be fought. Paying attention to *both* of these seemingly opposite processes is the kernel of the idea of natural history, to which we will now turn.

Natural History

How does the idea of the primacy of the object relate to nature? Answering this question is not easy, since Adorno uses the term "nature" in a variety of senses—depending, for instance, on whether he talks about the domination of nature, the stymied impulses of inner nature, or the transformation of society into a second nature.

To see what Adorno means by nature, we must begin with the idea of natural history—an idea that is fundamental to his works (as pointed out, for instance, by Cook 2011: 1, Pensky 2005: 228, Rose 1978: 38). He first formulated the idea in his 1932 lecture "The Idea of a Natural History," which drew on Benjamin's notion of natural history in *The Origin of the German Tragic Drama*, which was published in 1928.[15] Benjamin tried to show how history and nature were inextricably entwined in the German tragic plays of the Baroque. Nature was whatever appeared to be pregiven as fate, but even such nature was marked by history. The Baroque writers, according to Benjamin, did not see nature "in bud and bloom, but in the over-ripeness and decay of her creations. In nature they saw eternal transience, and here alone did the saturnine vision of this generation recognize history" (Benjamin 1985: 179). History was the constantly changing sphere of action, the stream of becoming, but earthly history was captured in blind fatefulness and thus in nature. In contrast to the days of divine creation, earthly history bore the marks of "the destructive effect of time, of inevitable transience, of the fall from the heights" (Benjamin 1985: 92). Benjamin pointed to the ruin, in which "history has physically merged into the setting" (Benjamin 1985: 178), as the embodiment of this disconsolate entwinement of nature and history. Seeing no hope in fallen earthly existence, the Baroque writers piled up ruins and fragments in the hope of a miracle. Earthly history was nature since it was fallen and unfree, but this nature was itself history since it was transient. Both were marked by the absence of divine truth.

In his lecture, Adorno condenses Benjamin's discussion into a sharp antithesis of history and nature in which the two terms dialectically revert into each other: nature is understood as historical and history as nature-like. The task, he writes, is:

> to grasp historical being itself, in its utmost historical determinacy, where it is most historical, as natural being, or if possible to grasp nature, where it seems most deeply and intractably natural, as historical being.
>
> (Adorno 1973b: 354f)

This sentence forms a *chiasmus*, a rhetorical figure that conjoins sentence parts that are like inverted echoes of each other. The formulation was so important for Adorno that he repeated it almost verbatim in *Negative Dialectics* more than three decades later (Adorno 1973a: 359).

More explicitly than Benjamin, Adorno uses the idea of natural history as a tool for criticizing reification. He lets the natural and historical relativize each other to break through the reification of each. This idea, he argues, offers an advantage over Lukács's treatment of second nature since it captures both directions of the dialectical movement—showing not only how the historical, supposedly the realm of the qualitatively new, has turned into a frozen image of timelessness, but also how what appears natural, as pregiven fate, contains traces that reveal it to be transient.

This explains the multiple ways Adorno uses the concept of nature. If nature and history revert into each other, neither can have a stable definition. Nature is never fully natural and history never fully historical. We must therefore apply the concepts in a provisional fashion and let them be undermined by the dialectical movement that is triggered when they encounter their objects. A helpful guide to the nuances that enter into the two concepts in the course of this procedure is offered by Susan Buck-Morss, who observes that Adorno typically employs both nature and history in a positive as well as a negative sense (Buck-Morss 1977: 54–9). Firstly, nature is sometimes used in a positive sense—for instance, when he discusses the vulnerability or mortality of living beings or when nature is seen as the victim of instrumental reason. Nature in this sense is almost always an object of domination. But nature is also, secondly, used in a negative sense, as a mythical world not penetrated by reason and hence outside human control. Whenever he refers to the Lukácsian idea of modern society as a second nature drained of meaning and unresponsive to needs, it is nature in this sense that is brought into play. This ambiguity of the concept of nature colors his notion of the primacy of the object, as Rolf Wiggershaus has noted. In a positive sense, this primacy means "openness on the part of a differentiated subject towards an object," but in a negative sense it means "domination over powerless individuals by social forces that had become autonomous" (Wiggershaus 1994: 602f).

History is likewise deployed in two senses. When used in a positive sense it stands for praxis that escapes pure identity and brings about the qualitatively new. When used in a negative sense, it stands for the reproduction of domination and identity, as seen in ideologies of progress or in the stage theory of orthodox Marxism. It is thus not the case that Adorno always favors history over nature. Invoking historical progress can be just as ideological as defending a social order as natural.

To complicate the four-part classification, no part can be isolated from the others since they are dialectically interrelated. The more we purify the category of nature, Adorno suggests, the more we are bound to discover historical elements in it, and vice versa. The point of the idea of natural history is to show that neither nature nor history can be absolutized. Both contain utopian as well as ideological aspects and are therefore ambivalent. Both turn bad the moment they are used to dismiss suffering and justify an oppressive social order. Nature stands for that with which reconciliation must be achieved but is also prone to ideologization when it presents historical creations as eternal and necessary. History is sometimes mythic or ideological, as in the notion of inevitable progress, but sometimes embodies a utopian hope, as in the idea of freedom.

The idea of natural history helps us see why Adorno's critical theory is neither fully realist nor fully constructivist. Viewing nature as an independent reality, in line with the idea of the primacy of the object, it insists that our representations of it are socially mediated. This explains why what Kate Soper has termed "nature-embracing" and "nature-sceptical" views of nature coexist in his writings. We both find attempts to *defend nature* against the encroachments of instrumental reason and a *critique of nature* as a reified semblance that hides its own historical origins. Both of these seemingly opposed views of nature also coexist in the public debate today, being backed, on the one hand, by environmental movements mobilizing to defend nature and, on the other hand, by feminist and anti-racist movements questioning and deconstructing notions of the natural. As Soper writes, an urgent task today for those who seek to defend nature is to search for a notion of nature that is not naively realist and that is viable in the face of decontructivist critique (Soper 1995: 8; compare Biro 2000: iv, Görg 2011: 51ff). Adorno, I suggest, makes important contributions to carrying out that task. If nature had been nothing more than a fluid category, as implied by constructivists, his thought would have had little to offer readers already steeped in postmodern relativism. By showing how thought, despite the fluidity of its categories, can nevertheless strive to do justice to the object, using the experience of pain or suffering to become aware of its own insufficiency, he points to a way beyond relativism.

Natural History and Second Nature

The idea of natural history brings out a significant difference between Adorno's and Lukács's concept of second nature, that is, society as perceived under the spell of reification. Adorno employs the concept of second nature as a moment of natural history, highlighting the mutual reversibility of history and nature into each other. This has two important consequences. Firstly, even first nature, when scrutinized, can be exposed as historically produced. This implies a constructivism that is less pronounced in Lukács. Even as he declares nature to be a social category, Lukács tends to rely on a static contrast between first and second nature and to view only the latter as reified and as devoid of life and meaning.[16] To Adorno, by contrast, all nature, including first nature, is history that has been *made* to appear ahistorical. His endorsement of a quote by Paul Valéry expresses the constructivist side of his thought: "There is no such thing as nature … [T]rees, the sea, the sun itself – and above all the human eye – all are 'artificial', in the last analysis" (quoted in Adorno 1991: 141).

But, secondly, the idea of natural history denies this constructivism. If nature returns in history itself, it is a delusion to think that humanity is free to shape its world as it pleases. Just as there is no pure nature, there is no pure history either. While Lukács's theory of reification is a valuable tool for analyzing how capitalist society congeals into a new nature that constrains humanity's ability to shape its history, his reliance on the idea of the proletariat as a historical macro-subject capable of overcoming reification through its own agency risks reproducing a reified philosophy of history premised on the possibility of a pure history. Adorno criticized Lukács because he worried that the latter overemphasized the capacity of the proletariat to shape history at the expense of nature and that this led to a rationalization of suffering—a worry that was aggravated by Lukács's treatment of nature as a mere social category.

To Adorno, the point of criticizing second nature is not to reawaken history at any cost. Attention also had to be paid to the vulnerability of the first nature which Adorno thought that Benjamin brought out in his discussion of decay and suffering.[17] To understand why Adorno here sides with first nature against second nature we must remember that the idea of natural history operates conjuncturally: the point is to criticize nature where it is *most* nature-like and history where it is *most* history-like. Just as he never simply champions history against nature, he never simply defends first nature against the second—but first nature should be defended wherever history or second nature appears stronger. Unlike Lukács, Adorno can therefore resort to nature to criticize reification. Since reified forms are imposed on nature as well as to history, simply reawakening history in the form of the human agency of the proletariat is insufficient. Prioritizing the human subject risks reproducing the harm done on nature. In the terms of the *Dialectic of Enlightenment*—which I will discuss below—what is needed isn't just more enlightenment, but an enlightenment that is enlightened about its self-destructive tendencies and the harm it inflicts on nature.

But can nature really be used to counteract reification? Isn't the point of stressing the harm done on first nature to appeal to *human* agency? Adorno's answer would be that such agency must be informed by a remembrance of nature, not least the inner nature that human beings suppress within themselves. Human beings are not themselves entirely subject but are also, in part, object. As he and Horkheimer point out in the *Dialectic of Enlightenment*, the fear of nature is also a fear of the nature within human beings themselves. A remembrance of this nature can be a counterforce to reification, even if nature itself cannot (Adorno & Horkheimer 1997: 254f). Any human action to stop reification must be conducted in tandem with an appreciation for the primacy of the object, for the non-identical in inner as well as external nature that is never wholly captured by human thought.

The core of the idea of natural history is that neither history nor nature should be absolutized. It is the recognition that concepts are never identical to their objects that enables us to see how history reverts into nature and vice versa. This shows why the idea of natural history cannot be severed from that of the primacy of the object. It is when we try to respect the object that we realize why our concepts and categorizations can never be absolute. The objects that we associate with nature can subvert our concepts of nature, forcing us to discover historical elements in them, and in the same way we can discover nature in the objects we view as historical. Insisting on the primacy of the objects means that the urge to define concepts like nature and history in a fixed way must be subordinated to the imperative of not capturing the object one-sidedly. This explains how the constructivist and anti-constructivists sides of the idea of natural history can be reconciled. The primacy of the object calls for and makes sense of both.

Natural History and the *Dialectic of Enlightenment*

With the help of the idea of natural history, we are in a better position to understand the *Dialectic of Enlightenment*, which Adorno and Horkheimer published in 1944 while in exile in the United States and which has become a classical point of reference for the Frankfurt School idea of a domination of nature. This work is commonly read as a pessimistic lament on how enlightenment reason reverts into myth and becomes a means of domination. Although several passages exist that appear to support such a reading, I prefer a different one.[18] Viewing this work as a pessimistic philosophy of history does not make much sense given the authors' intentions. A philosophy of history would fall prey to Adorno's own strictures on idealism by casting a conceptual net over history and reifying it. I will here argue that reading the *Dialectic of Enlightenment* in the context of the idea of natural history enables us to go beyond the stereotypical reading of this work as a bleak philosophy of history.

Let us have a look at the introductory chapter—which, as its title indicates, is on the *concept* of enlightenment, not its history. The chapter describes enlightenment reason as a self-preservative reason rooted in an archaic fear of nature. Reason helps humanity break out of myth, a timeless world not penetrated by reason, and dominate nature. However, since this domination requires people to suppress their own inner nature, "a denial of nature in the human being" is the price for this control (Horkheimer & Adorno 2002: 42). Enlightenment reason reverts into myth since humanity itself is subjected in the same instrumental, dominating fashion as external nature. As a result, society becomes a new or second nature, as mythical and fearsome as the first. Ironically, a fully enlightened world represents the victory of nature since it remains under the spell of self-preservation.

Based on the reading that the book simply substitutes the enlightenment belief in progress for an account of regression, the work has been subjected to repeated criticism. Doesn't the totalizing critique of reason leave the authors in an aporetic position, leading to political paralysis?[19] What is the ground for their intervention if they indict reason as such? In indicting reason as such, don't they engage in a transhistorical mode of argumentation that neglects the role of capitalism and class-conflict in generating the domination of nature?

A good place to start a defense of the *Dialectic of Enlightenment* is the concise formulation that "[m]yth is already enlightenment, and enlightenment reverts to mythology" (Horkheimer & Adorno 2002: xviii). As is readily apparent, the statement replicates the *chiasmus* of natural history. It is also a template for the book itself. The first half-sentence is developed in the chapter on the Odyssey, which demonstrates the sprouting of enlightenment thinking in the cunning used by the Homeric hero to triumph over the personified forces of nature. The second half-sentence is expanded in the chapter on Kant, de Sade, and Nietzsche, which shows how modern enlightenment turns mythic in the guise of an instrumental reason that no longer obeys limitations.

Recognizing the chiasmic structure of the work should caution us against a one-sided reading of the work as a fatalistic lament over the unidirectional increase in the domination of nature. As in the case of natural history, the *chiasmus* expresses a contradictory object. We should recall that the introductory chapter is explicitly about the *concept* of enlightenment. Rather than describing a historical development, it presents us with the contradictions inherent in the concept of enlightenment as the legacy of a history that contains both progressive and catastrophic elements. The authors are not claiming that enlightenment thought was progressive in Homeric times but regressive today. What they claim is that *both* the progressive and regressive sides of the enlightenment are part of the concept that we have of it. What they do in the book is therefore, in Hegelian fashion, to follow the constitutive moments of the concept through its mediations in order to present its aporetic, contradictory essence. This is done by showing how seeming opposites—myth and reason—revert into each other as each one of them proves itself insufficient in the face of the contradictory reality.

By reading the *Dialectic of Enlightenment* in the light of natural history, we can avoid seeing it as a totalizing critique that deprives the authors of a rational ground for their own intervention. It is perfectly possible for the authors to harbor hope in a self-reflexive reason, an enlightenment that is enlightened about itself. Rather than rejecting reason, they pin their hopes on an enlightenment that reflexively becomes aware of its self-destructive tendencies and that by recognizing itself as nature may free itself from the compulsive urge to dominate nature (Horkheimer & Adorno 2002: xviii). Using the *chiasmus* as model for understanding the work enables us to see it as open-ended regarding the direction of history. Rather than being a pessimistic philosophy of history, it becomes a critique of the very possibility of a philosophy of history and thus holds open the possibility of change. This should be borne in mind when we consider the book's treatment of nature. It shows, firstly, that the thesis of a domination of nature implies neither a total control over nature nor an irreversible increase in domination. Secondly, it shows that the nature that the book defends is not the static nature of myth. Rather, the nature in which reason may recognize itself and free from domination is one that goes together with history, with a reason shorn of its own mythic tendencies.

From the Slingshot to the Megaton Bomb

That the *Dialectic of Enlightenment* might give the impression of an idealist philosophy of history is understandable considering that it is only later, in *Negative Dialectics*, that Adorno spells out the philosophical moves that clarify the materialist content of the *Dialectic of Enlightenment*. It is in the latter work that the idea of the primacy of the object is explained and that the idea of natural history, earlier known only through the relatively obscure 1932 lecture, is presented to a wider audience.

It may be objected that my interpretation runs counter to Adorno's well-known statement in *Negative Dialectics* that although there is no universal history that leads from savagery to humanitarianism, "there is one leading from the slingshot to the megaton bomb" (Adorno 1973a: 320). Let us look at where the sentence occurs:

> Universal history must be construed and denied. After the catastrophes that have happened, and in view of the catastrophes to come, it would be cynical to say that a plan for a better world is manifested in history and unites it. Not to be denied for that reason, however, is the unity that cements the discontinuous, chaotically splintered moments and phases of history – the unity of the control of nature, progressing to the rule over men, and finally to that over men's inner nature. No universal history leads from savagery to humanitarianism, but there is one leading from the slingshot to the megaton bomb ... It is the horror that verifies Hegel and stands him on his head ... The world spirit ... would have to be defined as permanent catastrophe.
>
> (Adorno 1973a: 320)

Doesn't this passage imply that Adorno after all operates with something like a pessimistic philosophy of history? I would suggest that it instead exposes what history looks like from the vantage point of the *failure* of such philosophies. Adorno doesn't simply invert the ideology of progress into one of regression. While history as progress aligns itself with the subjectively meaningful, history as permanent catastrophe cannot be fully grasped from the standpoint of the subject. An asymmetry exists between the ideology of progress, in which hopes and aspirations are borne along with a history that redeems and fulfills them, and the idea of permanent catastrophe, in which hopes and aspirations are continually crushed. The latter confronts the subject with the power of objective forces, with meaningless adversity that shatters what is subjectively meaningful. Unlike a continuity of progress, a continuity of catastrophes can only be imagined as a jarring discontinuity, a unity of "chaotically splintered moments."[20]

That the permanent catastrophe cannot be fit into the idea of an idealistic philosophy of history also means that there is no fatalism to it.[21] As Adorno states in his essay on "Progress," the idea of progress contains a promise that cannot be jettisoned, namely, that the catastrophe will one day end—"that things will finally get better, that people will at last be able to breathe a sigh of relief" (Adorno 2005: 144).

> Progress means: to step out of the magic spell, even out of the spell of progress that is itself nature, in that humanity becomes aware of its own inbred nature and brings to a halt the domination it exacts upon nature ... In this way it could be said that progress occurs where it ends.
>
> (Adorno 2005: 150)

The surface impression that the *Dialectic of Enlightenment* presents us with a pessimistic philosophy of history should not obscure the central idea, which I believe is contained in the *chiasmus* about myth reverting into enlightenment and vice versa. Rather than serving as the foundation for a philosophy of history, its point is to criticize and undermine such philosophies and open our eyes to the possibilities for stopping the catastrophe.

Natural History as Critique

The idea of natural history is useful as a tool for both critique and perception.[22] Let me start with critique. At first sight, erasing absolute boundaries between nature and history seems to invite relativism and leave criticism rudderless. But that is only the case if the idea of natural history is asserted in the abstract. To be dialectical, the reversals between history and nature must emerge out of the encounter with the object, as determinate negations in which thought realizes that the reversal is necessary to do justice to the object. Unlike in idealist versions of the dialectic, there is no purely conceptual automatism or necessity that drives these reversals. Instead it is the relation to the object that must reveal why any concept, such as history or nature, is insufficient.

The ideas of natural history and the primacy of the object are intimately linked. It is because categories like nature and history are never identical to the object that we realize that they must move beyond themselves when we apply them to experiences. It is this movement that illuminates the object. When we discover that reality is never purely natural or historical we should therefore not resort to abstractions (like "hybridity") that desensitize us to the object. Better is to insist on the one-sidedness of concepts and let their failure spur them to develop. Thoughts, Adorno asserts, should be pushed to extremes. "The dialectic advances by way of extremes, driving thoughts with the utmost consequentiality to the point where they turn back on themselves, instead of qualifying them" (1978: 86). That is why, as he states in a provocative statement, "only exaggeration per se today can be the medium of truth" (2005: 99).

A problem with abstraction is that it removes us from the context in which concepts are applied. Adorno's strategy can instead be described as conjunctural, as an attempt to bring concepts as close to context as possible. Several commentators (e.g., Cassegard 2007: 39f, Gunster 2011: 213f, 220f,

Vogel 1996) have remarked that Adorno appears to be inconsistent when he criticizes second nature but not first nature: if society has become a new, second nature, then why not acknowledge that it possesses both benign and malign aspects, just as first nature? But this inconsistency disappears if we recall that for Adorno criticism should be conjunctural and sensitive to context: history should be mobilized against nature where the latter is powerful and vice versa.[23] Natural history is consistently employed as a tool for *criticizing the stronger*. It is precisely where the mythic semblance of second nature is strongest—namely, capitalist society—that criticism must bring out the harm done of first nature and side with it, and with history, against second nature.

The conjunctural nature of the critique explains why the *Dialectic of Enlightenment* stresses the reversal of enlightenment into myth much more than the opposite process—which in turn explains why the book may give the impression of pessimism. Today's world is one in which the reversal of enlightenment into myth is much more pronounced than the opposite process. Far from being premised on a pessimistic philosophy of history and political quietism, the work deliberately overemphasizes the predominant tendency in order to criticize it. Far from leaving the authors no room for action, their black portrayal of modernity is itself a critical intervention that includes conscious exaggeration. It is in this light we should read Adorno's invocation of the almost totally integrated society. Žižek points out that this invocation amounts to presenting a feared future outcome as a fact that has already arrived, precisely in order to stimulate us to avert it:

> While traditional Marxism enjoined us to engage and act in order to bring about the necessity (of communism), Adorno and Horkheimer projected themselves into the final catastrophic outcome perceived as fixed (the advent of the "administered society" of total manipulation and the end of subjectivity) in order to stimulate us to act against this outcome in our present.
> (Žižek 2008: 460)

As the best theoretician of this kind of intervention, Žižek points to the philosopher Jean-Pierre Dupuy, who has put forward the idea of "enlightened doom-saying" and post-apocalyptic retroactivity as a way of handling climate change and other environmental threats (Dupuy 2007/2008, 2013: 27f, 46f). Paradoxically, according to Dupuy, what prevents people from staving off the doom toward which the world is heading is their belief in their own free will and their ability to "act otherwise," which makes it hard to conceive of the doom as real. The solution, he argues, is to project oneself into the catastrophic future and look back from there at the present, thereby endowing the doom with retrospective reality. One must allow "the mind to project itself into the aftermath of the catastrophe, and treat the event in the future perfect tense" (Dupuy 2013: 204). Only by renouncing the seemingly hopeful prognosis that the catastrophe can be averted can we free ourselves from paralysis and act. This strategy can be observed in certain recent forms of environmentalism. Whereas environmentalism since the 1960s has usually rested satisfied with an apocalyptic rhetoric, pointing to the threat of a future global catastrophe as a possibility that can still be averted, Dark Mountain and similar groups that have sprung up in recent years exemplify what I call a post-apocalyptic form of environmentalism according to which the catastrophe is ongoing or is unavoidable (Cassegård & Thörn 2018).[24] "Curiously enough," one of the founders of Dark Mountain states, "accepting this reality brings about not despair, as some have suggested, but a great sense of hope. Once we stop pretending that the impossible can happen, we are released to think seriously about the future" (quoted in Cassegård & Thörn 2018: 570). These groups show that the giving up hope may be a way to gain hope. Sean Parson (2017) advocates making use of this paradox in activism, citing Adorno as the great theoretician of "revolutionary pessimism" in which politics is generated by "embracing doom."

Wilderness and the City

The idea of natural history also sharpens our perception of things that fall outside established categories. It can open our eyes to how seemingly natural things undergo historical change while artificial or historical creations, like technologies or cities, can appear natural. This in turn helps us understand political processes better, such as the mobilization of social movements. Political conflicts associated with nature can appear in the context of artificial environments, as in the case of urban environmentalism. Conversely, political conflicts associated with history can appear in the context of nature, as in the case of climate justice activism.

Let me use the term "historization" for the process whereby natural things are experienced as historical. Here's the environmentalist Bill McKibben:

> If the waves crash up against the beach, eroding dunes and destroying homes, it is not the awesome power of Mother Nature. It is the awesome power of Mother Nature as altered by the awesome power of man.
>
> (McKibben 1989: 60)

The sensation voiced by McKibben is one reason for the seeming obsolescence today of the so-called "wilderness cult" (Guha & Martinez-Alier 1997, Martinez-Alier 2003) that has traditionally been prominent in the environmentalism of the Global North.[25] From the point of view of the idea of natural history, it is easy to denounce the idolization of the wilderness as ideological, since all nature is naturalized history. Whether asserted naïvely in the romantic longing of people to escape artificiality or cynically marketized in the tourist industry, the idea of an unsullied wilderness masks social functions that belie its purity. To a large extent today's pressing environmental issues concern historicized natures, like the climate. An even more important reason for the mounting criticism of the wilderness cult is the strength of voices mobilizing under the banner of justice, often from indigenous groups. These groups are not concerned with preserving wilderness, but with maintaining a nature that is central to their livelihood and their sense of community. To them, the nature that is to be preserved is already mixed with historical elements, with a human presence and cultural significance neglected by wilderness lovers.

While it's tempting to read an endorsement of first nature into the *Dialectic of Enlightenment*, with its powerful denunciation of the domination over nature, Adorno's defense of first nature—both the external nature ravaged by technology and the realm of mute, preconscious inner impulses repressed by identity thinking—is always accompanied by a criticism of the ideological uses to which the idea of nature is put and should therefore not be understood as an expression of a wilderness cult.[26] This is brought out well in the chapter on natural beauty in *Aesthetic Theory*, where he extols art over natural beauty since the latter remains wedded to myth. The romantic idealization of natural beauty is ideological since such beauty is perceived only by those for whom nature is no longer an overpowering presence and who can afford to longingly gaze back at it. It is therefore a product of the society from which it is imagined offering escape.

The naïve appreciation of natural beauty masks its social function, presenting as immediate and spontaneous what is in fact mediated (Adorno 1997: 65).

To Adorno, nature's promise of happiness must ultimately be carried out, not by nature but by art. Art brings into nature an awareness of the scars that humanity inflicts on nature. "Consciousness does justice to the experience of nature only when … it incorporates nature's wounds" (Adorno 1997: 68). These scars bring with them an awareness of history. They also help us see that a truth content exists in the wilderness cult. Being a victim of human action, nature is experienced as a vanishing entity that can only subsist in perverted forms in a false society. In a capitalist society bent on exploiting it, nature can only exist at all by assuming the ideological functions available to it, such as providing industrial resources or consoling human souls with the pretense that it is still unsullied. That is one of its scars. The wilderness cult expresses sadness at the demise of things non-functional to us—whether they are physically destroyed or simply functionalized by being incorporated into capitalism.

The opposite of historization is naturalization, the process whereby the artificial takes on features associated with nature.[27] For Benjamin as well as for Adorno, the naturalization of capitalist society is not only a metaphor but also a matter of phenomenology. Big cities have become a new natural environment, with their own temporal rhythms and complex ecology. As the opposition between history and nature breaks down, the city takes on the features of untamed, terrifying nature: it becomes an environment dominated by the sensation of shock, as Benjamin famously showed in his essays (e.g., 1997a). Yet not only the terrifying but also the enchanting aspects of the first nature return in the city. Adorno points out that human products often congeal into gentle, idyllic versions of nature, such as the cultivated landscape (Adorno 1997: 70). More than Adorno, however, Benjamin had an eye for the city as a site of fascination.

> Not to find one's way in a city may well be uninteresting and banal. It requires ignorance – nothing more. But to lose oneself in a city – as one loses oneself in a forest – that calls for quite a different schooling. Then, signboards and street names, passers-by, roofs, kiosks, or bars must speak to the wanderer like a cracking twig under his feet in the forest, like the startling call of a bittern in the distance, like the sudden stillness of a clearing with a lily standing erect at its centre.
>
> (Benjamin 1997b: 298)

But with these enchanting aspects, Benjamin points out, mythical elements come into play—"a new dream-filled sleep came over Europe, and, through it, a reactivation of mythic forces" (Benjamin 1999: 391)—that strengthens the semblance that capitalism is timeless and eternal. This aspect of second nature, to which Lukács pays scant attention, helps explain why myth is so intractable and why the proletarian revolution that he hoped for failed to displace it. The idea of natural history offers a way to see beyond this mythical semblance by bringing out the transience of all nature. While the fragility and vulnerability of first nature is plain to see, it helps us detect how second nature too decays and passes away.

Although neither Benjamin nor Adorno discusses it, the rise of new struggles in the new nature of the city testifies to the reappearance of history in second nature. When people mobilize to protest against the bad water they are forced to drink or to protect a park or a neighborhood from demolition, this is not different in principle from when front-line communities mobilize to protect a river or to prevent a pipeline from being drawn over their ancestral land.[28] The environment that environmentalists mobilize to protect can be a product of second as well as first nature. In such struggles, history in the good sense of people experiencing themselves as capable of changing society may reemerge amid the bad nature that modern society has become.

Concluding Remarks—and the Problem of Capitalism

Adorno's idea of the primacy of the object is a cornerstone of critical materialism and the equally important idea of natural history functions as a critical tool for relativizing the boundary between the categories of nature and history. These two ideas are interconnected, since it is the primacy that Adorno accords to the object that underlies the dialectical reversals between the categories of nature and history. Together, they yield a materialism that is sensitive to the fact that nature is never completely within our grasp. It is a materialism that, instead of incorporating matter into a conceptual system, knows that conceptual systems are secondary to the objects.

That Adorno took decisive steps toward developing such a materialism is one of his core contributions to a critical theory of nature. There is, however, a serious charge against him and Horkheimer that must be addressed before I can conclude this chapter. The charge is that they engage in a transhistorical or anthropological argument that misses the role of capitalism and class. Martin Jay argues that instead of class struggle they focus on "the larger conflict between man and nature both without and within, a conflict whose origins went back to before capitalism and whose continuation, indeed intensification, appeared likely after capitalism would end" (Jay 1973: 256). In other words, they see domination as an anthropological constant rather than as specific to capitalism. This criticism has recently been reiterated by Adrian Wilding. While calling for a critical theory of nature building on the work of the Frankfurt School, he criticizes Horkheimer and Adorno for relying on an ahistorical argument about the human species that obfuscates the role of class and the divide between the global North and South. "Antagonism and struggle ... are curiously absent from *Dialectic of Enlightenment*," he writes (Wilding 2008: 54), pointing out that the authors rely solely on humanity's drive for self-preservation to explain the domination of nature while the latter is in fact the result of struggles within society.[29]

It is true that discussions of class and capitalism are nearly absent in the *Dialectic of Enlightenment*.[30] At the same time, three separate arguments can be made in partial defense of Adorno and Horkheimer. Firstly, it might be a sign of sanity that they see domination outside capitalism as well.[31] Even if the scale and pace of environmental destruction have increased manifold with capitalism, pre-capitalist humanity too exploited and transformed nature, pushing animal species into extinction by hunting and habitat deprivation. Moreover, Adorno may be right that the ingrained social forms of domination will not cease as long as the domination of nature is not changed. In "a seriously liberated vision of society that includes the relationship between man and nature, the relation to the domination of nature has to be changed if it is not constantly to reproduce itself in the internal forms of society" (Adorno 2008: 59). It has to be admitted, however, that this first line of defense only goes that far, since it is with the arrival of modern capitalism, and especially the so-called great acceleration of recent decades, that the destruction of nature reaches truly staggering proportions. This is a strong argument for focusing critical energies on capitalism rather than instrumental reason in general.

This brings us to the second partial defense, which is that class to a certain extent *does* play a role in the *Dialectic of Enlightenment* and that references to capitalism as well as contradictions in capitalist society are easier to find in other texts by Adorno, where he argues that it is only in capitalism that nature is subordinated to a relentless pursuit of surplus value cloaked in a principle of equal exchange (Adorno 2005: 159).[32] The domination of nature may well be a transhistorical process, but capitalism intensifies it by incorporating it into the peculiar dynamic generated by the commodity form and justifying it through the ideology of progress.

The third partial defense is that even if Adorno's argument is transhistorical it isn't anthropological in the sense of presupposing an ahistorical human nature. As I've argued, the argument of the *Dialectic of Enlightenment* should be read in the light of the idea of natural history. This also applies to his discussion of "idealism as rage" in *Negative Dialectics* that is repeatedly referred to by those who criticize him for relying on an anthropological framework.[33] The idealist conceptual system, Adorno writes, "has its primal history in the pre-mental, the animal life of the species." Just as hungry predators supplement their hunger with "rage at the victim," humanity justifies its drive to dominate and subsume the non-identical by projection. The idealist system is "the belly turned mind, and rage is the mark of each and every idealism" (Adorno 1973a: 22f).

These remarks may seem like a blatant example of a crude naturalism, a brutal reduction of mind to biology, but if we insert them into the idea of natural history we see that this naturalism only represents *one* side of the nature-history dialectic, the other side of which is that the belly is itself a historical creation. This means that the "idealism as rage" argument doesn't imply that Adorno resorts to ahistorical anthropological constants. Naturalism and historicism are meant to relativize each other. Just as the *Dialectic of Enlightenment* isn't a philosophy of history, the idea of "idealism as rage" isn't an anthropology.

If my interpretation is correct, no political quietism results from Adorno's position. As mentioned above, the charge that Adorno is making an anthropological argument is linked to the criticism that he ends up in a political deadlock. While the *Dialectic of Enlightenment* largely disregards class struggle, his work is still grounded in the contradictions generated by what he diagnosed as "late capitalism" (Adorno 1987). Although seldom manifested in open class struggle, they are evident in experiences of pain, shock, and despair. What is missed by those who claim that his writings end up in political withdrawal is that the totalizing picture he paints is the means whereby he hopes to achieve a critical effect.

A Critique of Concepts or of Capitalism?

Even if the charge that Adorno obscures the role of capitalism by stressing anthropological constants can be rebuffed, it is still possible to argue that his negative dialectics is primarily geared toward a critique of the relation between concepts and objects, inviting the question how effective it is as a critique of capitalism. How much does the relation between concepts and objects tell us about the relation between capitalism and nature?

Adorno clearly believes that translating the criticism of idealism into a criticism of capitalism is possible since capitalism behaves like an idealist system: it tries to grasp the essence of the world through its categories, aggressively subsuming the non-identical—whether human beings or other parts of nature. As he points out, Marx too understood the critique of political economy as *both* a critique of liberal economic theory and of the economy itself (Adorno 2018: 158).

However, isn't Adorno overlooking important differences between capitalism and idealism? Isn't the reduction of existing capitalism to those formal features that make it resemble idealist systems premised on a disregard of the object—a reductionist operation of the kind that his negative dialectics is designed to forbid? Isn't the messy reality of capitalism obscured when we view it as adequately captured by the concepts used to describe it? However, two arguments can be advanced in defense of Adorno. Firstly, institutionalized systems like capitalism do as a matter of fact rely on concepts to function. Concepts like value are embedded in practices that are constitutive of capitalism, forcing people to make use of them whether they regard them as credible or not. Institutions are therefore as oppressive of objects as purely conceptual systems. Hence it makes sense to criticize them in analogy with the criticism of idealism. The second argument consists in pointing out that Adorno employs an immanent critique that confronts concepts with the reality that they claim to represent. Seeing capitalism as a conceptual system therefore doesn't mean disregarding this reality. On the contrary, it means criticizing it by pointing to its inability of grasping the reality—an approach that we also see in Bloch when he discusses the anxiety of the engineer or in Hegel when he discusses measure.

Regardless of the validity of the extrapolation from conceptual to institutional systems, the relations between concepts and their objects and between institutions and their objects are *both* important to critical theory. For that reason, we should distinguish between *conceptual* systems and *institutional* ones, like capitalism, clearer than Adorno does. What he should be criticized for is not the extrapolation itself, but for a lack of precision concerning when he criticizes one rather than the other. This lack of precision makes the shape of the system that he criticizes nebulous. It can range from a philosophical system to the entirety of the modern world grasped as an integrated system. As we will see in next chapter, it is only with later thinkers—especially with Schmidt and the development of the so-called new Marx-reading—that a Frankfurt School critical materialism oriented more specifically to a critique of *capitalism* emerges.

4

Capitalism and the Domination of Nature

The human species has had a drastic impact on the environment ever since the start of the Holocene, but a truly catastrophic trajectory characterized by a rapid and accelerating rate of destruction becomes visible only with the growth of capitalism. Capitalism must therefore be given a special place in any theory that wants to come to grips with today's planetary ecological catastrophe. As we have seen, however, classical Frankfurt School treatises like the *Dialectic of Enlightenment* have been criticized for neglecting capitalism in favor of seemingly transhistorical arguments about instrumental reason and the domination of nature.[1]

Here a reconnect to Marx is needed. Alfred Schmidt takes important step in this direction, firstly, through his groundbreaking study *The Concept of Nature in Marx*, and secondly, through his role in the developments in the 1970s and 1980s from which the so-called new Marx-reading emerged. Like other thinkers who participated in these developments, Schmidt takes central insights from Adorno's negative dialectics but unhinges them from transhistorical claims in order to reconnect to capitalism. Schmidt, in turn, may have stimulated Adorno to redirect attention to capitalism in *Negative Dialectics* and other late writings from the 1960s (O'Kane 2015: 192).

Below, I start by introducing *The Concept of Nature in Marx*, focusing on a part of Schmidt's argument that has been most subject to much debate, namely, his seemingly "Promethean" embrace of the necessity of a continued struggle with nature—even in a socialist future.[2] The criticism against Schmidt is, I will show, largely based on misunderstandings stemming from a failure to appreciate his adherence to the methodological approach developed by his critical theory mentors Horkheimer and Adorno, in which, as he admitted, every page of his work was impregnated (Schmidt 2014: 9). Nevertheless, the criticism does correctly identify a weak spot in Schmidt, namely, his failure to clarify how a post-capitalist society might be less environmentally destructive than a capitalist one. This weakness reflects a tendency in Schmidt to pull back from a clear delimitation of critical attention to capitalism and instead see the domination of nature as rooted in metabolism as such.

This means that although the critique of domination becomes much more centered on capitalism than in Horkheimer and Adorno, two new and interconnected problems appear in his work. One concerns the delimitation of capitalism and to what extent environmental destruction reflects a specifically capitalist logic rather than transhistorical elements also present in non-capitalist societies. The other is the general problem of how to envision of a more reconciled relation to nature from the standpoint of a critical materialism focused on a criticism of the present—the problem of utopia.

Schmidt and *The Concept of Nature in Marx*

Schmidt's *The Concept of Nature in Marx* originated as a 1960 dissertation written for Horkheimer and Adorno and was published in 1962. It was a seminal work, providing the first systematic treatise on Marx's concept of nature and reviving interest in his concept of metabolism, the exchange of matter through which use-values are obtained from nature. Three points of Schmidt's Marx-interpretation deserve special mention: the sociohistorical concept of nature, second nature, and metabolism.

Firstly, in contrast to Engels's attempt to develop a dialectic equally applicable to nature and society, Schmidt stresses the socio-historical character of Marx's concept of nature (e.g., 2014: 10, 15f). In contrast to the objectivistic dialectics developed by Engels, Marx saw both nature and society as products of collective praxis. Nature existed in relation to human activity, the two being impossible to extricate from each other. Humanity was part of nature, and by changing nature humanity also transformed itself.

Secondly, a continuity with previous critical theory is evident in the weight Schmidt accords to the concept of second nature, the semblance of lawlike objectivity in capitalist society. Following Horkheimer and Adorno, he argues that humankind's domination of nature has been achieved in parallel to its own subjection to second nature, with the result that this domination is not experienced as liberating but as a blind, natural process beyond human control. In capitalism "the control of nature … remains at the same time an utter subjection to nature" (2014: 42). More than previous critical theorists, however, Schmidt turns to Marx to demonstrate how this second nature impacts on first nature: it is by clarifying the transformation of the economy into a reified system that we can understand how it mobilizes the use-values of first nature in the service of capital.

Thirdly, in a pioneering move Schmidt seizes on Marx's concept of metabolism to give a more concrete account than Horkheimer and Adorno of the domination of nature achieved in capitalism. As Marx points out in the first volume of *Capital*, all societies depend on a metabolism with nature—a continuous exchange of matter and energy that is mediated by concrete labor and that essentially concerns the movement of use-values: "it is an eternal natural necessity which mediates the metabolism between man and nature" (Marx 1990: 133).[3] In Schmidt's hands, the concept of metabolism drives home the importance of what Adorno calls the primacy of the object. Through metabolism, humanity is forced to interact with an alien element that will never conform exactly to its concepts and which therefore marks "the 'natural' limits of all historical dialectics" (Schmidt 2014: 11). Schmidt uses the term "negative ontology" to refer to the irreducible element of objectivity in nature that can never be assimilated to the categories of the subject, even though we can experience it through our practical engagement with it (Schmidt 2014: 86). This idea of negative ontology is central to Schmidt and shows that even though he may appear influenced by practical materialism through his emphasis on praxis, he differs from it by stressing that nature cannot be interwoven into a harmonious dialectical totality. This last point leads over to his infamous pessimism regarding the socialist future. According to Schmidt, even a society that supersedes capitalism must struggle with nature no less than other societies: "historically the incompatibility of man with nature, i.e. in the last analysis the necessity of labour, triumphs over the unity of man and nature" (Schmidt 2014: 30).

The Rejection of Engels's Dialectics of Nature

A prime target of Schmidt's criticism in *The Concept of Nature in Marx* is the conception of objective dialectical laws operating in nature independently of human praxis. In particular, he singles out Engels's so-called dialectical materialism for incisive criticism. He sees such objectivistic philosophies as implying a reification of history that must be rejected in a properly dialectical approach. The kernel of his criticism is expressed in a passage that concludes his criticism of Engels:

> Hence, it is only the process of knowing nature which can be dialectical, not nature itself. Nature for itself is devoid of any negativity. Negativity only emerges in nature with the working Subject. A dialectical relation is only possible *between* man and nature. In view of Engels's objectivism, in itself already undialectical, the question whether nature's laws of motion are mechanical or dialectical is distinctly scholastic.
>
> (Schmidt 2014: 195)

In contrast to the conception of dialectical laws operating in nature independently of humans, Schmidt points out that Marx himself held fast to a sociohistorical view of nature as existing in dialectical interplay with human labor and human praxis. It was nature in that sense that was the basis of his materialism, not the reified conception of objective laws.

The fact that Marx admitted of no fundamental methodological distinction between natural science and historical science didn't mean that he wanted to subject both to an objectivist metaphysics; instead, it meant that both had to be related to human praxis. By recognizing that nature is socially mediated, he avoided ontologizing it. Engels by contrast "relapsed into a dogmatic metaphysic" (Schmidt 2014: 51). Schmidt argues that Marx and Engels turned to positive science in quite different ways. Engels tended to a metaphysical materialism, Marx to a concept of nature as concretely mediated through society. This difference was mirrored in different conceptions of dialectics. Whereas to Engels, dialectics was a descriptive tool reflecting objective laws in nature that operated independently of humans, to Marx dialectics existed only in the relation between nature and humanity. Not objective laws but practice constituted dialectics. To Marx, there could be no question of "a dialectic of external nature, independent of men" (Schmidt 2014: 59). This also meant that, to Marx, the dialectic didn't admit of predictions in the manner of natural science: "The materialist dialectic is non-teleological," as Schmidt argues (Schmidt 2014: 35).

Schmidt's criticism of Engels has an obvious political addressee, namely, the orthodox Marxism of the Soviet Union. Engels's attempt to codify the dialectic by treating dialectical laws as natural laws prepared the way for the Stalinist dogma of the absolute objectivity of historical laws (Schmidt 2014: 192). Marx's project, Schmidt asserts, was different and much more emancipatory.

> [W]hen Marx wrote of the "natural laws" of society … this had the critical meaning that men are subjected to a system of material conditions which is outside their control and triumphs over them as a "second nature" … While Marx wanted these laws to *vanish* through being dissolved by the rational actions of liberated individuals, Engels naturalistically identified the laws of man within those of physical nature.
>
> (Schmidt 2014: 191f)

Maintaining a critical eye on both capitalism and Soviet-style communism was typical of Frankfurt School critical theory. In particular, the influence of Adorno is palpable in the affirmation of the vision of an incurable non-identity between humanity and nature—a non-identity that, Schmidt points out, remains even in socialism.

The Persistence of Non-identity

At this point, we might wonder how Schmidt's sociohistorical view of nature as existing in dialectical interplay with human labor relates to critical materialism. We can raise two questions. Firstly, while his criticism of Engels makes it clear that his stance is incompatible with *causal* materialism, does it really differ from *practical* materialism? Isn't the stress on praxis as mediator of humanity and nature precisely what distinguishes practical materialism? Second, what happens in Schmidt's oeuvre to Adorno's idea of nature as existing independently of human thought and work? Isn't the primacy of the object jettisoned if nature is viewed exclusively as a product of praxis and labor?

The simple answer to both questions is that Schmidt *is* a critical materialist whose stance is clearly different from that of practical materialists like Lukács or Bloch. I have already pointed out how he, in line with Adorno's idea of the primacy of the object, uses Marx's idea of metabolism to point to the irreducible otherness of nature.[4] Despite its close association with labor and praxis, nature is always viewed by Schmidt as non-identical and irreducible to social categories. Many passages illustrate the deep imprint on his thinking of Adorno's idea of non-identity.[5] Schmidt stresses that to Marx, nature is "that which is not particular to the Subject" (Schmidt 2014: 27). Nature therefore retains the otherness from thought which Adorno tried to capture with his idea of non-identity. While Hegel's thought is in unity with itself, "[i]n the Marxist dialectic the reverse is the case: it is non-identity which is victorious" (Schmidt 2014: 28). Following Adorno, Schmidt therefore criticizes Lukács for a tendency to dissolve nature into social forms, pointing out that "in Marx nature is not *merely* a social category" (Schmidt 2014: 70). Bloch is criticized for advocating the idea of a telos of history, which goes against materialist non-identity and leads to reifying nature as the bearer of a superhuman idealist subject which is just "nature-speculation" (Schmidt 2014: 37, 59, 160f).

The side of Schmidt's thought that stresses non-identity is most evident when he criticizes romantic conceptions of nature. In contrast to the romantic idea that overcoming alienation must mean the resurrection of a full unity between subject and object, he affirms a model of reconciliation with nature borrowed from Adorno which is modeled on the recognition of an incurable non-identity between humanity and nature.[6] Even if the abstract labor associated with capitalism can be superseded, concrete labor and the necessity for a continued struggle with nature will not (Schmidt 2014: 11). This means "that classless humanity will also be confronted with something ultimately non-identical with itself" (Schmidt 2014: 86). Due to this non-identity, labor will always require the suppression of instincts (Schmidt 2014: 137). Nature will never be "completely 'made' by us," "even in a truly human world there is no full reconciliation of Subject and Object" (Schmidt 2014: 158). Finally, "nature's co-production with labour always includes the fact that what men have in mind always remains utterly foreign and external to it. Even under socialism" (Schmidt 2014: 162).

Schmidt thus *both* asserts that nature exists only in interplay with human praxis and labor and that nature retains an indelible otherness that makes it irreducible to human subjectivity. The question is how he combines these two assertions. The answer is the idea of metabolism. Humans are inescapably bound to an exchange with a nature that eludes the full grasp of human consciousness. Nature becomes dialectical through this exchange with human acting subject but always retains an alien element that cannot be reduced to the intentions of these subjects—a fact that Schmidt referred to as negative ontology.[7]

Schmidt's rejection of the idea that the end of capitalism will also end the non-identity between humankind and nature has invited controversy. Even in socialism, he writes, people will still have to struggle with nature.

> With socialism, nature's objectivity does not simply disappear ... but remains something external, to be appropriated. In other words, men will always have to work.
>
> (Schmidt 2014: 71)

> Since the realm of necessity will continue to exist as long as human history, men will always be compelled to behave towards nature in an essentially appropriative, interfering, struggling manner.
>
> (Schmidt 2014: 157)

Unlike the young Marx of the Paris Manuscripts, the mature Marx of *Capital* realized that socialism would not resolve the antagonism between humankind and nature:

> In later life [Marx] no longer wrote of a "resurrection" of the whole of nature ... Nature is to be mastered with gigantic technological aids, and the smallest possible expenditure of time and labour ... When Marx and Engels complain about the unholy plundering of nature, they are not concerned with nature itself but with considerations of economic utility.
>
> (Schmidt 2014: 155)

In a passage alluding to Engels's famous warning about the "revenge" of nature, he writes:

> The exploitation of nature will not cease in the future but man's encroachments into nature will be rationalized ... In this way, nature will be robbed step by step of the possibility of revenging itself on men for their victories over it.
>
> (Schmidt 2014: 155f)

I will return to these controversial lines shortly, but already here we should note two things. The first is that, provocative as the lines may seem today, Schmidt is not unfaithful to Marx, who in volume 3 of *Capital* asserted:

> Just as the savage must wrestle with nature to satisfy his needs, to maintain and reproduce his life, so must civilized man, and he must do so in all forms of society and under all possible modes of production ... Freedom, in this sphere, can consist only in this, that socialized man, the associated producers, govern the human metabolism with nature in a rational way, bringing it under their collective control instead of being governed by it as a blind power; accomplishing it with the least expenditure of energy and in conditions most worthy and appropriate for their human nature. But it always remains a realm of necessity. The true realm of freedom, the development of human powers as an end in itself, begins beyond it, though it can only flourish with this realm of necessity as its basis. The reduction of the working day is the basic prerequisite.
>
> (Marx 1991: 958f)

The second thing to note is that Schmidt, despite his rejection of romantic utopias of a restored unity with nature, is not proposing that a continuation of the industrialist exploitation of nature we have seen in capitalism is inevitable. The following passage hints at the peace that would characterize a more reconciled relation to nature:

> What could be salvaged from the idea of such a very naive relation to nature ... is the hope that when men are no longer led by their form of society to regard each other primarily from the point of view of economic advantage, they will be able to restore to external things something of their independence, their "reality" in Brecht's sense. In such a society, men's view of natural things would lose its tenseness, it would have something of the rest and composure which surrounds the word "nature" in Spinoza.
>
> (Schmidt 2014: 158)

I will return later to how I believe that passages like this, with their faint utopian luster, can be reconciled with the grim formulations quoted above about the need for a continued struggle with nature in a post-capitalist society. Before doing that, I will turn to the eco-Marxist criticism of Schmidt, which has seized on his seeming endorsement of a ruthless exploitation of nature.

Burkett's Critique

That eco-Marxists are critical of Schmidt is not surprising. Since they want to show Marx's usefulness for a green, environmental politics, any Marx-interpretation that presents him as insensitive to nature is bound to come in for criticism. In a harshly dismissive review of Schmidt's work, Paul Burkett claims that Schmidt inadequately analyses the contradictory and exploitative character of value and capital. Firstly, he overlooks the possibility that capitalism will generate an ecological crisis that will undermine capitalism itself—the fact that "capitalism is the first society capable of a truly planetary catastrophe due to a tendency to 'undermine the conditions of its own exploitation.'" Secondly, "Schmidt's interpretation falsely undermines certain grounds for pro-ecological working-class politics" (Burkett 1997: 174). Resorting to heavier artillery, Burkett adds: "In Schmidt's hands, Marx' vision of communism becomes an anti-ecological industrial utopia" (Burkett 1997: 165). Furthermore, Schmidt's interpretation "lapses into an uncritical determinism similar to that of official (Stalinist) Marxism. This determinism unjustifiably naturalizes capitalism's exploitation of nature while bypassing the systemic basis for an eventual merging of Red and Green anticapitalist movements" (Burkett 1997: 164).

This is a remarkable misreading. The first thing that strikes one as odd is Burkett's seamless assimilation of Schmidt into "official (Stalinist) Marxism." This assimilation recurs in several places in Burkett's text, for instance when he writes that "Schmidt's analysis ... encapsulates the best and the worst of official Marxism's stance on Marx's environmental implications" (Burkett 1997: 164). But such an assimilation is clearly incorrect considering that Schmidt's entire book is dedicated to an attack on such "official Marxism" (for a few passages that explicitly criticize such Marxism, see e.g. Schmidt 2014, 39f, 108).

Burkett misses that Schmidt's dialectical approach, which was mainstream in critical theory when he wrote his book, has an inbuilt allergy to determinism of the kind associated with Soviet-style "official Marxism." Accusing Schmidt of "lapsing into an uncritical determinism" is a fundamental and glaring misreading considering that Schmidt himself explicitly rejects all teleology in Marx (see e.g. Schmidt 2014: 35, 1981: 15, 19). That history is not determined by objective laws but by praxis is a message that Schmidt repeatedly emphasizes as central in Marx. To claim, as Burkett does, that Schmidt's account is "similar to official Marxism's objectivist emphasis on desocialized productive forces as the motor of social evolution" (Burkett 1997: 165) is clearly a gross error.

While it is true that Schmidt is suspicious of all ideas of a resurrected unity of humankind and nature, this is not because he affirms industrialism. Furthermore, it is certainly *not* because he is an adherent of "official Marxism." The reason is rather to be found in his adherence to Adorno's philosophy of non-identity: he is issuing a warning that post-capitalist society will not bring about perfect unity or identity in the relation to nature, and that any reconciliation to be had must build on relinquishing the drive to identity. Schmidt's insistence on the continued need for a struggle with nature after capitalism is not due to adoration of Soviet-style industrialism, but to his skepticism of identity. When Burkett mistakes Schmidt's rejection of the romantic idea of wholeness for an embrace of Soviet-style industrialism, he is seeing Stalin's shadow where he ought to see Adorno's.

Burkett is also incorrect in his claim that Schmidt *affirms* the continuing struggle with nature. He correctly quotes Schmidt's statement that even a future socialist society would exploit nature (and maliciously adds: "a projection similar to that promulgated by Stalin," Burkett 1997: 170). But what he fails to tell us is that Schmidt *never claims to endorse or celebrate* this vision of the future. Indeed, Schmidt's view is unmistakably bleak, almost Weberian. To call this akin to Stalinism or official Marxism is widely off the mark.

The reason for this misreading of Schmidt must be sought in the rather complex position the latter adopts. As I have shown, Schmidt follows Adorno in arguing *both* for a socio-historical conception of nature as existing in interplay with human praxis *and* for a primacy of the object in the sense that nature possesses an indelible otherness or non-identity, which means that it can never be reduced to human subjectivity. Two things appear to have caused

Burkett's misreading. Firstly, he overlooks that the thrust of the former of these two ideas—that of the socio-historical conception of nature—goes against the grain of any objectivistic dialectics of nature as exemplified by Engels or official Soviet-style Marxism. Secondly, an even graver error is that he misunderstands Schmidt's critique of the romantic ideal of a restored unity with nature, which was directed against idealist philosophy and certain Western Marxists such as Bloch, as an endorsement of ruthless industrialization and exploitation of nature. Burkett's criticism is grounded in a failure to grasp the two-thronged nature of Schmidt's criticism, which is directed *both* against objectivistic approaches to nature and against romantic dreams of unity. As a result, he also misses the real contribution of Schmidt, namely, his attempt to stake out a position that avoids these opposite pitfalls by building on Adorno's ideas of natural history and the primacy of the object.

Burkett's misreading of Schmidt thus seems connected to a failure to grasp the particular form of dialectics as practiced by the latter, which clearly bears the mark of Adorno's negative dialectics. This suspicion is strengthened when we turn to the criticism of Schmidt by his close colleague Foster (Foster 2016a, Foster & Clark 2016a, 2016b). To be sure, Foster's criticism is more nuanced and careful than Burkett's. Foster sensibly makes no attempt to assimilate Schmidt into Soviet-style "official Marxism" and he correctly observes that Schmidt's book is infused with Weberian-style pessimism (Foster & Clark 2016b: 103). In the main, however, Foster follows Burkett, reproducing several of the latter's misunderstandings. He thus portrays Schmidt as a determinist (Foster & Clark 2016a), despite Schmidt's oeuvre being based on a rejection of the determinism of Engels and subsequent Soviet-style Marxism. He also claims that Schmidt misses "the full significance of Marx's historically specific critique of the capitalist value form, in which value, emanating from labor alone, was in contradiction to wealth, deriving from both nature and labor" (Foster & Clark 2016a), which is a strange and at least highly unnuanced assertion since Schmidt explicitly acknowledges that nature contributes to use-value but not to value or exchange-value (Schmidt 2014: 65f). Furthermore, he claims that Schmidt attributes to Marx a conception of nature as "passive and mechanical" (Foster & Clark 2016a), implying that Schmidt reads a Cartesian dualism into Marx, which again is not true since Schmidt explicitly states that to Marx humans are part of nature and human labor power is a force of nature (Schmidt 2014: 16). He also follows Burkett's lament about Schmidt's failure to appreciate the ecological side of Marx and about his pessimism regarding the possibility of an end to the antagonism with nature—but theoretically his criticism reaches deeper than Burkett's, taking aim at the very conceptions of dialectics and materialism in the Frankfurt School. I will therefore deal with Foster's criticism separately. I will return to his eco-Marxist version of dialectics in Chapter 7 in order to point out its weaknesses and present a more general defense of the Frankfurt School conception of dialectics.

What Truth Is There in the Criticism?

Let me return to Schmidt's controversial statements about the necessity of a continuing struggle with nature. The fact that the eco-Marxist criticism of Schmidt is based on a series of misunderstandings doesn't mean that their dissatisfaction with his work is wholly groundless. What Schmidt should be criticized for, however, isn't embracing a Promethean Soviet-style industrialism, but refusing to come up with a clear explanation of how his own idea of a post-capitalist struggle with nature would differ from the Soviet variant. *This* is where Schmidt's weak point is, as he seems to have recognized himself by the time he added a new preface to *The Concept of Nature in Marx* in 1970. Echoing Benjamin's words that rather than blindly pursuing mastery over nature, a liberated humanity should aim for mastery over its *relation* to nature (Benjamin 1997b: 104), he states in this preface that what Marx wanted was not just a quantitative increase in man's mastery over nature, but:

> mastery by the whole of society of society's mastery over nature. This mastery would certainly still depend on the functions of instrumental reason. But since it would "finalize" these functions, and subject them to truly human aims … society's mastery over nature would thereby be freed from the curse of being simultaneously a mastery over men, and of thus perpetuating the reign of blind natural history.
>
> (Schmidt 2014: 13)

This clarification doesn't go very far, and it doesn't justify the more dramatic passages in the work itself where Schmidt seems to equate socialism with a drive to "master nature with gigantic technological aids" and so fully "rationalize" its domination over nature that nature is "robbed step by step of the possibility of revenging itself on men for their victories over it" (Schmidt 2014: 155f)—all formulations that suggest a kind of merciless warfare against nature. The fact that human beings must work, or that nature must remain a realm of necessity, doesn't mean that their relation to nature must be one of perennial antagonism or domination. Nothing in Schmidt's or Marx's arguments about metabolism justifies the idea that the relation to nature must take that form.

Another way to formulate the criticism against *The Concept of Nature in Marx* is that it tends to adhere to a transhistorical concept of labor that abstracts from the historical specificities of its embeddedness in capitalism. Adhering to a transhistorical concept of labor is not necessarily wrong, since *concrete labor*—unlike abstract labor—is indeed best understood as transhistorical. But by abstracting labor from the relations of production, Schmidt seems to neglect his own assertion that the relation to nature is always mediated by history (Smith 2010: 43f: 34, Yates 2018). Where he stands out in comparison with eco-Marxist critics like Burkett is in his tendency to blame environmental destruction not mainly on the rule of value and abstract labor, but rather on an inherent tendency toward domination in concrete labor as such and in the way such labor is employed in all metabolism with nature. What is thereby rendered invisible is how the domination of nature might take different forms in capitalism and in a post-capitalist future. While it is certainly true that human beings will always to some extent be engaged in concrete labor and therefore be forced to act instrumentally in relation to nature, it is only with the reign of value and abstract labor that nature is reified into a mere resource and a systematic *compulsion* arises for capitalists to intensify its utilization for the sake of capital accumulation. One should therefore not straightforwardly extrapolate the role played by concrete labor under the conditions of capitalism to a possible post-capitalist future.

Although the idea of a transhistorical process of unceasing domination of nature might seem to echo the argument of the *Dialectic of Enlightenment*,

Schmidt in fact breaks with that argument since Horkheimer and Adorno never claimed that domination would never end. Although they were as critical of the Soviet system as Schmidt was, they always took care to leave room for the utopian possibility of a reconciliation with nature. Moreover, one might argue that using Adorno's philosophy of non-identity to underpin an ahistorical vision of unceasing domination is perverse. Adorno never one-sidedly condemned only the romantic longing for wholeness and reconciliation. He was just as vehement in his condemnation of the opposite standpoint, that of modern instrumental reason suppressing inner and outer nature. Being locked in seeming opposition, these two standpoints were both reified, blind to the fact that they generated each other and constituted two sides of the same coin. In seeming to dismiss only the romantic pole of this opposition while stoically accepting the necessity of the instrumental pole, Schmidt takes leave of Adorno's more dialectical approach and becomes more Weberian than his mentor was in the sense of grimly acknowledging the unavoidability of a continuing domination of nature. While Schmidt's criticism of the idea of identity is due to Adorno, his bleak vision of socialism owes more to Weber.

In portraying socialism as striving for nothing more regarding nature than perfected domination, I believe that Schmidt is too un-nuanced. While his bleak portrayal may have been an apt description of the so-called actually existing socialism of the Soviet Union and its satellites, it fails to clarify why what Foster refers to as the metabolic rift *must* remain even in a post-capitalist world. The mere persistence of non-identity doesn't mean warfare, just as the necessity for work doesn't mean that the metabolic relation to nature must be one of domination. Schmidt appears to take the appropriateness of the label "socialism" for the Soviet system at face value. But thereby he forgets that critical theorists like Horkheimer and Adorno wanted a *different* and more genuine socialism. Forgetting this leads not just to the practical problem of how to conceive of a more reconciled relation to nature but also creates theoretical unclarity. To the extent that the capitalist compulsion to accumulate is seen as the root of the destructive warfare on nature, its continuation in a post-capitalist society is left unexplained. If this warfare is instead seen as an aspect of metabolism as such, then the delimitation of critical attention to capitalism is undermined and historical specificity is lost.

Feuerbach and the Turn to Ecology

Early signs that Schmidt self-critically starts to reevaluate some of his positions in *The Concept of Nature in Marx* can be seen in his treatise on Ludwig Feuerbach—*Emanzipatorische Sinnlichkeit: Ludwig Feuerbachs anthropologischer Materialismus*—which appeared in 1973. Here he shows an awareness of ecological problems that is lacking in *The Concept of Nature in Marx*, and he is clearly concerned to distance himself from the seeming insensitivity to nature in his earlier work. This can be seen in how he relativizes his earlier emphasis on labor as the primary form of activity that mediates our relation to nature. Earlier he had assented to Marx's criticism that Feuerbach adhered to an objectivist idea of pure nature existing per se, independently of the context of human praxis. Now, however, Schmidt is ready to defend Feuerbach's contemplative attitude. In revaluating Feuerbach, he also takes his first steps toward working out how a relation of non-identity to nature can be a basis for respecting otherness rather than for ruthless struggle. Aesthetics, he suggests, provides a model for a sensual-contemplative approach to nature that allows nature to appear to us as a subject in its own right rather than as an object to be subjected to instrumental reason. With approval, he quotes Feuerbach saying that while thought is intolerant, the senses let the object be what it is, namely, Subject: "Only the senses, only the contemplation [*Anschauung*] gives me something as subject" (quoted in Schmidt 1973: 31f). Whereas Schmidt had earlier contrasted his sociohistorical approach to nature with the contemplative attitude of modern science, he now discerns another, more benign form of contemplative approach to nature, namely, the passive, sensual-contemplative relation to natural things that leaves them in peace and acknowledges them as subjects (Schmidt 1973: 45, 47).

This move represents a significant modification of his earlier argument about the unavoidability of a struggling, antagonistic relation to nature. It also brings him closer to Adorno, in whose writings aesthetics and sensual experience occupied a key role in sensitizing thought to non-identity and in which mimesis came forward as a possible model for a reconciliation with nature. Instead of one-sidedly criticizing the romantic longing for wholeness and unity, this turn to aesthetics helps Schmidt develop a sharper criticism of rationalism and instrumental reason than in *The Concept of Nature in Marx*.

Ecological concerns are even more explicitly foregrounded in the new preface that he wrote to *The Concept of Nature in Marx* in 1993 (Schmidt 2016). Here he reconsiders the book in the light of the widespread awareness of ecological crisis and of the limits to growth. He provides a new, more nuanced discussion of Marx's ecological consciousness, which in many respects comes close to the glowing portrayal of Marx as an ecological thinker among eco-Marxists, although he highlights the complexity of Marx's position more than them. Self-critically, Schmidt remarks that the original book tended to discuss the human relation to nature almost exclusively from the perspective of the subject-object schema of labor and knowledge while neglecting other equally valid aspects of Marx. What he believes is missing is clarified through his appreciative words about Feuerbach's contemplative materialism, which he sees as a model for not violating nature, for perceiving it aesthetically rather than with practical egoism. Our material exchange with nature in production is only one of many possible ways of interacting with it. The historical dialectic, he asserts, must expand into an ecological materialism. Only today do we realize the extent to which society is encompassed in nature. That insight should caution us not to assume that socialism will result in human mastery over nature. Much would be won, he adds, if humanity rejected limitless growth and reoriented itself toward living in harmony with nature.

Schmidt's writings after *The Concept of Nature in Marx* thus evince a steady growth in ecological consciousness. This goes hand in hand with an increasing appreciation for thinkers like Feuerbach, whose appreciation of aesthetics as a model for a non-dominating relation to nature is needed as a supplement to Marx's and Engels's hopes for a post-revolutionary future in which nature would be subjected to rational control. What remains problematical, however, is that Schmidt's reorientation fails to affect his valuation of labor. As we have seen, *The Concept of Nature in Marx* recognizes that labor is historical in the sense that its abstract, value-producing aspect is limited to capitalism, but he nevertheless holds on to a negative view of labor, which even in its concrete aspect is portrayed as a form of warfare or struggle. This makes it hard for him to imagine how an end to the ecologically destructive relationship to nature can come about even with a transition to socialism. The revaluation of Feuerbach and the shift of emphasis from labor to aesthetic contemplation is no solution in this regard, since the metabolism with nature will continue to depend on concrete labor and since he nowhere makes any suggestion that aesthetics can fundamentally refigure such labor. In this sense, he remains faithful to Marx's assessment in *Capital* that labor will always remain a realm of necessity.

Toward the New Marx-reading

Schmidt contributed to the critical theory of nature not only through his explicit discussions about the relation between capitalism and nature, but also by participating in the reorientation of critical theory toward a closer reading of Marx together with other young scholars who had been active in the Frankfurt milieu around Adorno, such as Hans-Georg Backhaus and Helmut Reichelt, an approach that took off in the 1970s and led to the so-called new Marx-reading.[8] This is a clear attempt to read Marx through the lens of critical materialism—a materialism that is not transhistorical but directs its critical energies toward a totality limited in time, namely, capitalism.

By early on recognizing the limits of dialectics in investigating this encounter with matter, Schmidt's *The Concept of Nature in Marx* had already taken important steps in the direction of critical materialism. Building on Adorno's idea of the system as a negative totality in conflict with the objects to which its categories were applied, he recognized that analysis must deal with capitalism as a *limited* totality imposing itself as a second nature on society, and that critical theory had to show how this totality was dependent on a continuous process of metabolic exchange with the outside world. The rise of second nature is described as a process whereby society itself turns into natural environment, setting "its own essence against its creators" (Schmidt 2014: 16). Second nature is thus described, in Hegelian fashion, as an alienated totality that humanity can no longer recognize as its creation. But rather than expressing the essence of a genuine and inclusive totality in the Hegelian sense of the world as experienced in thought, it expresses the essence of a false or negative totality, which is partial even as it claims universality. Hegel had described first nature as "blind conceptless occurrence" while second nature was the world of men as it had consciously taken shape in the state and its laws. To this Schmidt opposed the view that "Hegel's 'second nature' should rather be described in the terms he applied to the first: namely, as the area of conceptlessness, where blind necessity and blind chance coincide" (Schmidt 2014: 43). To Schmidt, as to Adorno, the whole was the false.

This argument foreshadows the systematic treatment of Hegel in *History and Structure: An Essay on Hegelian-Marxist and Structural Theories of History*, a 1971 work in which Schmidt emphasizes the model of Hegel's *Logic* for Marx's exposition of his argument in *Capital*. To be sure, Schmidt doesn't overlook the crucial differences between the *Science of Logic* and *Capital*. Unlike Hegel, Marx limited his analysis to the totality constituted by capitalism. Concepts used in his exposition such as value or abstract labor therefore have no validity outside capitalism. Since this totality is negative it calls for critique.[9] As Schmidt points out, bourgeois society "blindly and violently establishes itself over against its creators" and the "rigidified, thinglike, congealed relationships between people ... form a negative totality, which, in turn, is the specific object of dialectical materialism" (Schmidt 1981: 44). Rather than being a mere exposition of the totality, the dialectic must therefore be an immanent critique of the totality. Referring to Adorno, Schmidt states that the laws of capitalist society should be *abolished*—neither disregarded as in overly voluntarist variants of Marxism nor extrapolated into a general theory of society, as in structural Marxism (Schmidt 1981: 81).

Schmidt's version of the new Marx-reading has many attractive features. If Adorno often seems to direct his critique equally at institutional and conceptual systems, Schmidt moves the center of analysis back to capitalism. This move also helps overcome the Frankfurt School's tendency to neglect political economy in favor of cultural analyses. Delimiting totality to capitalism furthermore clarifies how it is possible to assert with Adorno that "the whole is the false" without having to deny the existence of opposing forces—indeed, of brewing popular discontent—outside the totality. This ties in well with the central idea in *The Concept of Nature in Marx* about the contradiction between the capitalist second nature and the material reality on which it depends. Keeping this contradiction in full view is important to avoid the risk that the focus on presenting the constitutive forms of capitalism may lead to overstressing the harmonic unity and functionality of the negative totality.[10] However, a weak spot can be identified in Schmidt's position. Despite his attempt to delimit his critical attention to capital, Schmidt never really clarifies the ambiguity that exists in *The Concept of Nature in Marx* about whether environmental destruction is primarily to blame on capitalism or on inherent features of metabolism as such.

Conclusion and Remaining Problems

In this chapter I have followed Schmidt's development from *The Concept of Nature in Marx* to later writings where he both modifies his earlier arguments and expands them to cover the aesthetic contemplation of nature. I have highlighted Schmidt's contributions to the critical theory of nature—above all his pioneering study of Marx's concept of nature, his delineation of the negative totality to capitalism and his participation in developing the new Marx-reading. These contributions go a long way toward reconnecting critical materialism to Marx and rectifying Horkheimer and Adorno's neglect of the economy, but arguably they fail to bring sufficient light on capitalism's relation to nature. This shows up in two problems in his account.

The first problem concerns the relation between negative totality and its outside. His writings are ambiguous regarding to what extent the domination of nature is rooted in concrete labor, which is a necessary part of all metabolism with nature, and to what extent it results from a particularly capitalist logic connected to value and capital accumulation. He tends to portray concrete labor as inherently dominating, but it is also easy to find warnings in his writings against abstracting Marxian categories from capitalism. He thus writes that Marx's argumentation in *Capital* always presupposes the relation to capital, meaning that even seemingly anthropological statements relate to the bourgeois age, not humanity as such (Schmidt 1976: 64–8, 2014: 10). To remove the theoretical ambiguity concerning whether concrete labor unmediated by capital must take the form of a domination of nature an important remaining task is thus to clarify to what extent Marxian concepts, such as concrete labor, are transhistorically applicable. Without such a clarification, no clear picture of *how* capitalism ravages nature can emerge.

The second problem is related to the first. It concerns how to conceive of a more reconciled relation to nature in a post-capitalist society, that is, the problem of utopia. I have discussed the vehement criticism that Schmidt's views on this matter in *The Concept of Nature in Marx* have invited. Although these views are modified in his subsequent writings, as can be seen in his revaluation of Feuerbach's contemplative materialism, the implications of this revaluation for a possible reorganization of society are hardly spelled out. His texts after *The Concept of Nature in Marx* take him closer to a more ecological vision of a post-capitalist future but say little about how the metabolism with nature in such a future might be organized.

Both problems touch on the ability of critical materialism to theorize capitalism's outside, which includes nature as well as the possibility of non-capitalist relations in the future. A necessary first step to elucidating the problems is to work out more concretely where the boundary between capitalism and its outside should be drawn and how it functions as an interface regulating the metabolism between capitalism and nature. Despite the injunction to focus on the relation between first and second nature, Schmidt places the emphasis a little too much on the first, on metabolism. As he himself admits, his discussion in *The Concept of Nature in Marx* is mainly concerned with use-value and the natural form of the commodity (Schmidt 2014: 66). We therefore need to supplement his analysis by turning to the role of the value-form in structuring and regulating the interaction with nature. The point is not to choose between focusing on second and first nature or value and use-value, but having the tools needed for an analysis that illuminates the contradictory relation between them. I will turn to this task in the next chapter, where I will clarify the respective roles of concrete and abstract labor in the domination of nature with the help of Marx. There I will also discuss the question of the transhistorical applicability of concepts such as concrete labor. By doing so I will provide an answer to the first problem above. As for the second and more general problem of how a post-capitalist reconciliation with nature might look, I will return to that in the book's final chapter.

5

Marx, Value, and Nature

"Labor is *not the source* of all wealth", Marx writes in his critique of the Gotha Programme. "*Nature* is just as much the source of use values (and it is surely of such that material wealth consists!) as labor" (Marx 2010d: 81). The explosive power of this sentence becomes evident when read in conjunction with the so-called labor theory of value, which states that what counts as *value* in capitalism, as opposed to material wealth, is derived solely from labor. Typical of capitalist value is that it doesn't reflect the use-values provided by nature, even though these use-values belong to the preconditions of labor. Capitalism depends on nature but fails to value it. Nature thus forms part of a fundamental contradiction in capitalism, namely, between material wealth and value.

Exploring this contradiction is the subject of this chapter. The point of departure is the problem brought up in the previous chapter. There I argued that Alfred Schmidt's difficulty in formulating a post-capitalist utopia in *The Concept of Nature in Marx* is not just reflective of the general difficulty of thinking the outside of the criticized totality, but is more directly rooted in a failure to clarify how the value-form is directly responsible for the particularly destructive form that the metabolism with nature has taken in capitalist societies. The result of this failure is Schmidt's tendency to blame, not capitalism, but metabolism as such for this destructive relationship.

To rectify this lack of clarity regarding how capitalism ravages nature I will turn to Marx's own writings, in particular *Capital*. Although he focuses his attention on capitalism's relation to the proletariat, his labor theory of value is also highly enlightening regarding how capitalism regulates its exchange with nature. The value-form, I will argue, plays a crucial role in structuring capital's differential treatment of labor and nature.[1]

Below I start with a presentation of the labor theory of value. This will help us clarify nature's "outside" status vis-à-vis capitalism, namely, as an outside on which capitalism is crucially dependent but which is only *indirectly* integrated into the valorization process. This closer look at Marx's value law requires us to pose the question of how seemingly transhistorical categories like nature, wealth, concrete labor, and metabolism relate to his historically specific analysis of capitalism that is centered on concepts like abstract labor and value. In line with my critical materialist reading, I stress that Marx's intention in *Capital* is not to set up a transhistorical framework for theorizing history in general. That, however, doesn't make his use of transhistorical concepts an anomaly. Such concepts point beyond the negative totality and are mobilized to highlight the non-identity of that totality with the reality on which its categories are imposed. Transhistorical concepts like concrete labor and use-value therefore play a crucial role in Marx's critique of political economy.

I close the chapter by discussing more recent attempts to go beyond a pure value-form analysis that take their point of departure in the Marxian value law, thereby demonstrating the fruitfulness of his theory for further developments of the critical theory of nature.

Value, Labor, and Nature

Why is nature regarded as an outside of capitalism to being with? What prevents capitalism from including the nature on which it is, after all, dependent? Why does the borderline between them exist and what systemic features in capitalism contribute to reproducing it? To illuminate these questions, we need to turn to what Marx's value law says about capitalism.

In capitalism, the value of a commodity is an expression of socially necessary labor time, that is, the labor time that society on average needs to produce the commodity. However, although labor provides the *substance* of value, its *form*—the value-form—emerges from the market where the commodities are compared and exchanged, and where value becomes visible in the medium of money or exchange-value. Marx points out that the value-form is social and shouldn't be confused with the natural or physical form of the commodity as a bearer of use-value. Unlike the natural form, the value-form is what makes the commodity exchangeable with other commodities. "Not an atom of matter," he writes, "enters into the objectivity of commodities as values" (Marx 1990: 138f).

As can be seen, distinctions between the natural and the social are central to Marx's account of the value law. Commodities thus have a dual nature, possessing both a natural form as carriers of use-value and a social form, or value-form, that is mediated by the market. *Use-value* is the usefulness of a thing and doesn't depend on exchange. For useful things to possess *value*, they must turn into commodities that can be exchanged on the market—that is, they must acquire the value-form. *Use-values* exist in all societies, but in capitalism *value* is predominant since market exchange is the principal nexus for integrating and coordinating economic activities. Already here we see the possibility of the contradiction between wealth and value mentioned above, since many use-values never become commodities and thus remain unvalued in capitalism. Corresponding to the dual nature of commodities is that of capitalist labor, which is both concrete and abstract. *Abstract labor* is Marx's term for value-producing labor. It is typical of capitalism and should be kept analytically separate from the actual concrete labor carried out by workers engaged in production. *Concrete labor* exists in all societies and is expended every time we create a use-value. Like use-value, concrete labor is not necessarily mediated by the market. Abstract labor, by contrast, can be determined only through the market, since it is only when the commodity acquires the value-form that it is possible to identify the abstract labor needed to produce that value.[2]

The market's crucial mediating role in capitalist production gives rise to *commodity fetishism*—the appearance that value is an inherent property of the commodities themselves despite deriving its substance from labor.[3] Fetishism means that value is not simply an expression of labor time, but a *reified* expression that invisibilizes its own production process. Reification, as we recall, is the process whereby a thing appears to possess its properties independently of its context, as inherent or ahistorical attributes. In this invisibilization of material processes capitalism and idealism are similar. To Adorno, both are characterized by a rule of abstractions that in the case of capitalism stems from the act of exchange in which the exchanged products are reduced to an abstract equivalent. To use Alfred Sohn-Rethel's (1978) term, the abstractions are "real abstractions"—concepts that are part of capitalism's constitutive practices and therefore have material force seemingly independently of human intention.[4] As a result, the economy appears to people as an objective second nature, or, as Marx writes, "as overwhelming natural laws, governing them irrespective of their will" (Marx 1991: 969).

Fetishism, then, arises because value production includes a crucial moment—that of market exchange—that invisibilizes the role of labor in this process. To this, I want to add that nature too is invisibilized.[5] If I buy a table, it is not just labor but also the trees used for the wood, the sunlight that helped them grow and the material used for the tools that appear irrelevant to the table's value. Fetishism is therefore best understood as a *general* invisibilization of the preconditions underlying the production of value, including the contribution of nature. Marx hints at this himself when he describes the mystification brought about by fetishism as "the bewitched, distorted and upside-down world haunted by Monsieur le Capital and Madame la Terre, who are at the same time social characters and mere things" (Marx 1991: 969).

Here we should be aware of a hidden trap. If fetishism means that value is treated as independent of labor as well as nature, how about commodities like oil, land, or timber where the prevalent tendency is precisely to view value as inherent in the resource itself, in nature? Is it possible to claim that the contribution of nature to the value of such commodities is invisibilized? This objection, however, dissolves at a moment's reflection, because according to Marx it is only *labor* that produces value, not nature. Treating natural resources as inherently valuable is precisely what obscures their real contribution, which is that they are *use-values*.

Nature, Use-value, and Value

In opting for a broad definition of fetishism as invisibilizing both nature and labor, we should remember that while capitalism depends on both, the difference is that while labor produces value, nature does not. Nature nevertheless has an indispensable *indirect* role in the creation of value by providing use-values for capitalist production such as nutrients, raw material, and energy. Just as abstract labor provides the substance of a commodity's value-form, nature and concrete labor provide it with its natural or physical form.[6]

Grasping the intricate relation between use-value and value is crucial. To begin with, no commodity has value if it lacks use-value (since such a product wouldn't sell), and use-value springs from labor as well as nature—"labour is its father and the earth its mother," as Marx writes (1990: 134). If we look more specifically at how use-values affect value, we soon detect a significant difference between two fundamentally different kinds of use-value. First we have the use-value of labor power, which—in Marx's famous words—"possesses the peculiar property of being a source of value" (1990: 270). Being able to produce more value for the capitalists than is needed for its own reproduction, this is the secret of surplus extraction in capitalism. Again it should be stressed that it isn't the input of concrete labor per se that constitutes value. Instead the point is that the value of a commodity, as it is determined at the end of the circulation process when it is put on the market, exceeds the value that goes back to the worker in the form of wages.

Secondly, other use-values than labor power have a startlingly different effect on value. Paradoxically, such use-values generally *detract* from the value produced by labor when deployed in the production process. Access to plentiful natural resources or machinery tends to increase the productivity of labor, which reduces the amount of labor that is socially necessary for producing a given commodity and hence it also reduces the value of that commodity. What nevertheless spurs capitalists to increase productivity is competition. The striving for individual profit compels them to search out ever cheaper resources and ever more efficient technologies, even if it lowers the overall value produced in capitalism.

Here the question may arise how nature can be cheap or not if it lacks value to begin with. As Marx points out, even things without value can have a price. Nature can only be freely appropriated as long as it isn't turned into private property, but once that happens it ceases to be free except to its owner (1990: 197). The theory behind the pricing of nature is provided by Marx in his discussion of ground rent in the third volume of *Capital*. As he points out, the discussion of ground rent holds true for all natural resources that can be monopolized, for example, waterfalls, mines, fishing grounds, or well-situated building sites (1990, 1991: 908). When such property is commodified, a tendency to monopoly pricing sets in, meaning that prices are set by production costs and by what buyers are willing to pay rather than by competition. This means that nature is a realm where prices seldom reflect value. The price of oil or real estate, for instance, is less an expression of labor time than of monopoly pricing. This, however, doesn't invalidate the labor theory of value. As Paul Burkett points out, even natural resources that are monopolized and yield a rent are "still freely appropriated from the standpoint of capital as a social whole, since rents merely redistribute pre-existing values" (2014: xviii).

Is the Labor Theory of Value Anthropocentric?

A repeated accusation against Marx's labor theory of value is that it is anthropocentric since it claims that only labor—not nature—creates value in capitalism. As Burkett points out, this accusation is groundless. The value law does not reflect any normative bias in Marx's thinking, but the way capitalism works. It is not Marx, but capitalism that should be blamed for refusing to grant value to nature (Burkett 1999, 2014: xvif). This argument can also be found in Schmidt, who notes that the accusation against Marx for ecological insensitivity is unfair, since he neglects nature only because the bourgeois economy itself does this (Schmidt 1973: 34).

Nevertheless, the question may still be asked why capitalism treats labor power alone as a source of value in capitalism.[7] Why not technology, energy flows, or other forms of constant capital? At first glance, it seems obvious that robots and other machines are just as capable as human workers of producing a surplus compared to what is needed for their reproduction. To avoid the charge of anthropocentrism, shouldn't one rather admit that the crucial input that creates value is simply *energy*, whether this passes through human bodies or not?[8]

Although such an admission might be tempting, it would be a mistake. A thought experiment shows why.[9] If we want to produce a fixed amount of a commodity, it might seem that labor power (variable capital) and other forms of energy (constant capital) are equivalent since both simply add energy to the production process. From a physical point of view, the two sources of energy are indeed comparable. The difference, however, becomes apparent if we imagine a change in social conditions that creates incentives for capitalists to increase either of them. Introducing more or better machinery can be expected to increase productivity and prices will therefore tend to fall. In terms of the value law, the increasing energy flows lower the amount of socially necessary labor for producing the commodity and hence its value. The opposite effect follows if labor power is increased. Assuming that the commodity is sold and that wages remain constant, the increase in working-hours drives up the costs for labor and hence also prices. In terms of the value law, the result is a more valuable commodity.

Why does the increase in the flow of energy result in a less valuable commodity in one case and a more valuable one in the other? Since the decisive difference doesn't reside in physical properties, the only possible answer is that the *form* of wage labor is crucial, regardless of whether this labor power is provided by humans or non-humans. By the form of wage labor I mean that labor is treated as a commodity that is sold on a labor market as the private property of individual workers. Using Marx's terms, wage labor alone can become abstract labor, while other types of energy input are mere concrete labor—labor that is useful but fails to show up as value. What matters, then, isn't whether labor power is *human* or not, but whether it is *waged* or not.

The thought experiment helps us see why. Due to the form of wage labor, increasing the input of labor power for a fixed amount of a commodity must result in increasing production costs, unless workers are found who are willing to work for cheaper wages. Increasing other forms of energy input, by contrast, is usually done to *lower* production costs by decreasing the socially necessary labor needed for manufacturing the commodity. This difference regarding production costs means that labor power sets the baseline for value while other forms of energy subtract from value in proportion to their cheapness in relation to labor power.[10]

A consequence—which might sound fanciful but should be spelled out to illustrate the absence of anthropocentrism in the labor theory of value—is that it is quite possible to imagine robots or animals producing value if they are employed as wage workers. The crucial role of the wage is shown by the fact that, just as waged robots or animals can produce value, human workers can lose their value-creating ability if their labor isn't provided as individual wage labor. If, for example, a capitalist pays a lump sum to a sub-contractor who agrees to provide the capitalist with the labor power necessary for a certain task, then labor time can be increased without affecting production costs for the capitalist. The same holds for a labor force consisting of serfs producing for a capitalist market. The effect in both cases is similar to the effect of increasing productivity by using unwaged machinery.[11] In a nutshell, from the point of view of the value law, unwaged labor tends to behave like constant capital while waged machinery, if it were to exist, would tend to behave like variable capital.

That the form of wage-labor is crucial for the value law of course doesn't mean that individual capitalists must try to preserve or increase the element of wage-labor in production. On the contrary, while capitalism as a whole depends on the production of value, individual capitalists compete for a share of the aggregate amount of surplus-value by trying to increase productivity, and cutting wage costs. This is why they struggle to minimize their reliance on wage-labor by scanning the earth for cheaper natural resources and spurring the development of ever more efficient technologies. It is also why capitalism co-exists with a wide variety of irregular and often unwaged forms of labor. Although unwaged labor doesn't create value, it gives individual capitalists a competitive edge against other capitalists and helps them increase their share of the aggregate surplus-value produced in society.

Use-value and Material Wealth

By mentioning use-values we have already gestured to the outside of capitalism. Not all use-values become commodities: friendship, for instance, or sunlight or intact eco-systems. All these things contribute to what Marx calls *wealth*. Capitalism only systematically reproduces wealth to the extent that it takes the form of value.

Moishe Postone points out the implications of distinguishing between wealth and value for analyzing the environmental crisis. Since in capitalism value can be created only by labor, capitalism increasingly comes to be characterized by a contradiction between the processes generating value and wealth—between the expansion of labor on the one hand, and the sacrifice, on the other hand, of the kinds of wealth that fail to count as value in capitalism, including nature. The system of producing for the sake of surplus-value results in "the accelerating destruction of the natural environment" (Postone 1993: 24ff, 193–200).[12]

It is significant that Marx when analyzing capitalism finds it indispensable to refer to concepts such as use-value and wealth that point to a reality beyond this system. We can now begin to elucidate the role that these concepts play in his presentation. They are important not only for theoretical or descriptive purposes, but also as *critical* tools.[13] By gesturing to things outside the exchange situation, they point to the non-identity of things hidden behind the equality of value. Here we see the explanation for Marx's predilection for conceptual pairs like value and wealth, concrete and abstract labor, or use-value and exchange-value. Typically, each pair combines a transhistorical concept with one that is specific for capitalism. The function of the former is not to theorize the non-capitalist world, but to reveal a contradiction at the heart of capitalism, the dissonance that arises from the fact that objects do not fit in with the system's categories. *Not* to pay attention to wealth and use-value would certainly yield a more harmonious picture in tune with capitalism's own self-image, but such a picture would also be abstract and ideological.

Metabolism and Reified Nature

The contradiction highlighted by the binary of use-value and value is mirrored in the uneasy relation between two qualitatively different ways in which capitalism behaves toward nature. It both *reifies* nature by subordinating it to the logic of value-creation and depends on a constant *metabolism* with nature that allows it to appropriate use-values. Schmidt argues that a twin focus is therefore necessary, on both capitalist society's metabolism with *first* nature and its transformation into a reified *second* nature. In his view, this double basis for Marx's argumentation should always be kept in mind and we should always ask ourselves "if we are speaking about second or first nature" (Schmidt 1968b: 29).

An important point that emerges from thinking about these two processes in conjunction is that they do not necessarily run smoothly together. Although conjoined, they can also disrupt each other. To explain, let us look closer at the process of reification. Just as the rule of the value-form reduces people to the reified forms prepared by the system (Rubin 1973: 22f), it reduces nature to a resource, inert soulless matter. Marx describes this process vividly in the

Grundrisse, where he discusses how capitalism gives rise to a tendency both to objectify nature and to develop natural science:

> Hence exploration of all of nature in order to discover new, useful qualities in things ... The exploration of the earth in all directions ... the development, hence, of the natural sciences to their highest point ... For the first time, nature becomes purely an object for humankind, purely a matter of utility; ceases to be recognized as a power for itself; and the theoretical discovery of its autonomous laws appears merely as a ruse so as to subjugate it under human needs.
>
> (Marx 1993: 409f)

The incorporation of nature into capitalism in the form of a resources means, paradoxically, that even things that we are accustomed to thinking of as first nature—plants, soil, minerals, and so on—become part of second nature, the realm of objectivity generated by capitalism. Second nature is not purely manmade or social but arises through the reification of both society and nature. What characterizes second nature is not that it is wholly artificial, but that it is reified, meaning that it appears ahistorical and "thinglike" despite being historically shaped by capitalism.

Reification means that reality is made to conform to forms that are part of second nature. Form is not just a mere external container, but constitutive for the matter at hand. It is thanks to form that nature becomes what it is in capitalism, namely, a resource. This means that nature is not simply a material world outside capitalism, but actively *constituted as an outside* by capitalism itself. As Jason Moore (2015) points out, it is *as an outside* lacking value but possessing use-value that it is incorporated into the second nature created by capitalism. The separation between society and nature isn't itself natural but reproduced by capitalism.[14] Unlike in precapitalism, in which the earth was what Marx called man's inorganic body, in capitalism man is separated from the earth, which is objectified and made available for exploitation. Capitalism dissolves the precapitalist community, which was characterized by a "natural unity of labour with its material presuppositions" and in which the aim of work was the creation of use-values rather than of value (Marx 1993: 471).

But the reduction of nature to a resource has a material side, highlighted by the concept of metabolism, that shows that the process can be disrupted by physical processes outside the reified forms. Metabolism is not just an enabler of capitalism but also a source of friction. As Marx points out, capitalist agriculture "disturbs the metabolic interaction between man and the earth, i.e. it prevents the return to the soil of its constituent elements consumed by man in the form of food and clothing; hence it hinders the operation of the eternal natural condition for the lasting fertility of the soil" (Marx 1990: 637). Capitalism thus simultaneously lays waste to workers and to nature, thereby "undermining the original sources of all wealth – the soil and the worker" (Marx 1990: 638, see also 1991: 950). This is what Foster (2000) refers to as Marx's theory of the metabolic rift. Today, the negative effects of this rift are not limited to the soil but include an immense sacrifice of use-values in the form of pollution, depleted resources, and staggering negative effects on other lifeforms.

In an often-quoted passage in *Capital*, Marx claims that subsuming nature under capitalist forms and property relations is an absurdity comparable to slavery.

> From the standpoint of a higher socio-economic formation, the private property of particular individuals in the earth will appear just as absurd as the private property of one man in other men. Even an entire society, a nation, or all simultaneously existing societies taken together, are not the owners of the earth. They are simply its possessors, its beneficiaries, and have to bequeath it in an improved state to succeeding generations, as *boni patres familias*.
>
> (Marx 1991: 911)

That *Capital* is a critique of political economy clearly doesn't mean that Marx limits himself entirely to analyzing the dominant categories of capitalism. Concepts such as metabolism, use-value, concrete labor, and material wealth perform a crucial function in pointing beyond those categories. That they are transhistorical in scope doesn't indicate that he intends to conduct a transhistorical analysis of either nature or society. Instead, they are brought to bear on the analysis of capitalism precisely as part of a *critique* of the latter. They demonstrate the absurdity of the negative totality by confronting its petrified forms, which have turned into second nature, with the reality that they repress.

The Critical Use of Transhistorical Categories

The transhistorical range of certain of Marx's concepts has been subject to much debate. Reasonable textual grounds seem to me to exist for interpreting value and abstract labor as integral to capitalism while use-value, concrete labor, and wealth are concepts with a transhistorical scope.[15]

Critical materialism rejects the aspiration of scientific Marxism to create a general, abstract theory about history as such, but this doesn't mean that it must confine itself to interpreting all of Marx's concepts as strictly internal to capitalism. Yates (2018) argues that categories like use-value and labor should be understood as determinations of capital, not as transhistorical. To be sure, Marx does use them as determinations of capital, but he never claims that use-value or concrete labor is *not* transhistorical. In fact, he explicitly claims that they *are* transhistorical—concrete labor, for instance, is unambiguously described as "an eternal natural necessity which mediates the metabolism between man and nature" (Marx 1990: 133; see also Marx 1990: 290)—although he at the same time mobilizes them as determinations of capital. This is perfectly understandable if one recalls Hegel. In his *Science of Logic*, he describes the positive judgment (e.g., "the rose is fragrant") as a mutual determination of subject and predicate (the individual is the universal and vice versa). The subject is the concrete individual with many qualities (hence universal), while the predicate is the abstract universal (which isolates the qualities and hence becomes individual). Hegel points out that "the predicate is determined in the subject; for it is not a determination *in general*, but *of the subject*" (Hegel 1969: 633f; compare Hegel 1969: 661f). In the same way, seemingly transhistorical abstract universals (predicates) such as use-value or concrete labor are brought in by Marx as determinations of capital. They remain universals (transhistorical) but are nevertheless determinations *of capital* (the subject).

Why is this important? It shows that Hegel and Marx can analyze totalities concretely *without having to reduce their concepts entirely to the totality* under investigation. Instead, a sense of non-identity is maintained between the abstract predicates and their individualized function within the totality. In other words, what Adorno calls non-identity is introduced in their thinking.

The subject and predicate are mutually determining, but a tension remains between them. Hence, such predicates become points of exteriority from which to criticize the totality. It is precisely because use-value and concrete labor are not fully internal to capitalism that they can be used to criticize the logic of capital. That is why I believe it is a serious mistake to focus *only* on categories that are fully immanent to capitalism. Doing so would yield a too harmonious picture of capitalism by removing its contradictions.

To return to Schmidt, I have argued that a weakness appears in his account due to his tendency to see domination as inherent in concrete labor as such. What is problematic is not that he uses a transhistorical notion of concrete labor, but rather that he fails to clarify that transhistorical notions are never sufficient to grasp anything concrete. They must always be applied in conjunction with other concepts to assume their proper, historically specific meaning. That means that the concrete forms assumed by labor can never be carried over or extrapolated from one historical period to the next.

We can now see the function of Marx's transhistorical concepts clearer. Even though they are transhistorical, they are brought to bear on historically specific, or concrete, situations. This means that they are often conjoined with concepts specific to capitalism. As mentioned, we see this in Marx's fondness for conceptual pairs that often serve to bring out contradictions. The dual nature of labor, for instance, refers to the conjunction in labor of social form (value, abstract labor) with a material and technical process of production (use-value, concrete labor). Concepts like use-value and concrete labor are essential to critique since they gesture toward the material side of this duality. This, I believe, is the reason that Marx is unwilling to tie such concepts exclusively to capitalism. Such concepts point to the object with which the system's categories must be confronted.

To no small part, the critical effect of *Capital* stems from the skillful use of juxtapositions in which categories of the system are confronted with accounts of how the system affects its outside. This explains why Marx devotes so much attention to the working day and the struggles around it. The reason is not that toil and subjugation are important categories of the system, but that he wants to *criticize* the system. The possibility of class struggle depends on the idea of a clash, or experienced non-identity, between the categories of capital and the lived experience of workers. Class struggle doesn't spring just from the fact that the standpoints of labor and capital are opposed, but also from workers' experience of themselves as *more* than labor, as non-identical to labor.

Going beyond Value-Form Analysis

A value-form analysis brings out the logic behind capitalism's different treatment of labor and nature clearly. But this clarity exists only as long as we grasp capital as an abstract model, or ideal type. Marx's model of the logic of capital is a theoretical construct, the logic of which hardly ever imposes itself on objects in pure form. As soon as we approach capitalism as a lived experience, we notice that it is embedded in contexts that blur its contours. If we accept, for instance, that patriarchy and imperialism are to a large extent sustained and reproduced due to their functional interrelationship with capitalism, then we must also recognize that capitalism can appear in forms such as sexism and racism that at first sight have little to do with economic categories.

This matters since experience is the very touchstone for critique in the critical theory of nature. It is in experience that Adorno seeks the spark of non-identity that might ignite a critical impulse—the sense of discrepancy that is perhaps at first only manifested in a vague unease or in the sensation that "something's missing" (Adorno & Bloch 1988: 1ff). That the capitalist logic is always embedded in a variety of lifeworlds means that nothing guarantees that a theoretically defined contradiction such as the one between value and wealth coincides with the site of the experience of non-identity. Depending on the context, the category sensed to do violence to non-identity may just as well be one of race or gender as of capital. This means that while we should certainly have a clear sense of capitalism as the negative totality to be criticized, it still makes sense also to analyze traces of instrumental reason and domination in life-contexts more generally, as Horkheimer and Adorno did. What is needed isn't to choose between analyzing the economy and seemingly non-economic domains, but to illuminate both in their interconnection.

If experience is a touchstone for critique, then we shouldn't just consider the abstract logic of capital but also the institutional contexts in which it is embedded. A recent example of how to do this is provided by the political philosopher and critical theorist Nancy Fraser. While building on Marx, she at the same time shows why we often encounter capitalism not directly but indirectly through nature, the family, or state power. Marx's *Capital*, she explains, performs two epistemic shifts. The first takes us from the publicly visible façade of capitalism, the equality of market exchange, to the "hidden abode" of production where the exploitation of wage labor is revealed. The section on primitive accumulation performs a second shift that brings us to an even more hidden realm, an "abode behind the abode," namely the realm of expropriation (Fraser & Jaeggi 2018: 40). As in Marx's classical account of primitive accumulation, expropriation is about the how capital accumulation is furthered through means other than the exploitation of wage labor in the strict sense. Fraser argues that capitalism depends on three realms that provide background conditions for capital accumulation, namely, non-human nature, the state, and social reproduction. She refers to them as backstories of capitalism—necessary to it but banished from the foreground of the official economy. Importantly, capitalism tends to destabilize these realms by ruining nature, emasculating political regulation, and monetizing human relations and thereby undermining the trust needed for social reproduction. This gives rise to what she calls boundary struggles that concern capitalism's destructive relation to its external conditions of existence (Fraser & Jaeggi 2018: 90–3, 150–4).

Suggestive as Fraser's theory is, it has been criticized for lacking a firm anchoring in Marx's theory of value (Saito 2017b). The discussion in this chapter about how use-values contribute indirectly to value by serving as preconditions for value-creation brings out more precisely how expropriation furthers capital accumulation. Expropriation is not necessarily about transfers of value (although it can be that as well, when it involves the non-market transfer of still tradable commodities produced by labor). More often it simply involves the incorporation of use-values into the capitalist production process, which they cheapen and facilitate. While becoming more important for capital in times of crises—gaining access to new sources of energy and other resources being a clear example—expropriation is also an everyday necessity for capitalism since it relies on the use-values provided by nature as well as by unwaged human labor, such as household work. Even if such use-values do not contribute directly to value, and in many cases lower the value of the commodity by increasing labor productivity, the pressure of competition in capitalism creates a systematic pressure for intensifying their use. The pressure explains why capitalism continually needs to deplete nature, in addition to exploiting wage-labor.

Fraser's sketch suggests some directions in which we may need to go beyond a pure value-form analysis in order to explore how the capitalist logic interrelates with nature and other seemingly non-economic domains that provide use-values for the accumulation of capital. Her notion of backstories usefully captures the ambiguity of domains that simultaneously appear to be both outside and inside capitalism—outside by being defined as non-economic conditions of possibility for capitalism, yet inside since they are relied upon by capitalism in order to facilitate capital accumulation. Her sketch, however, is only one example of how processes of expropriation can be related to the logic of the value-form. Within the field of Marxist scholarship, world-systems theory in particular has long stressed the importance of such processes and backed it up with considerable historical research. In Chapter 8, we will see how Jason Moore, inspired by such a perspective, again takes up the distinction between exploitation and expropriation (or appropriation, as he calls it), highlighting the role of "cheap nature" in capital accumulation.

Concluding Remarks

The aim of this chapter has been to use Marx—and in particular his account of how the value-form creates a systematic pressure in capitalism for a differential treatment of labor and nature—to illuminate the complex functioning of the boundary or interface separating capitalism from nature. Schmidt, as we have seen, tends to see domination as an inherent feature of concrete labor as such, and hence of all human metabolism with nature. In contrast to his account, I have stressed that while concrete labor partakes in the domination of nature in capitalist societies through its association with the value-form, nothing says that this role must continue in a post-capitalist society. Concrete labor contributes to capital accumulation by providing use-values that *indirectly* contribute to value but this role is hardly inherent in concrete labor as such. Although all concrete labor of course involves physical energy—and thereby by necessity depletes natural resources at least in the minimal sense specified by the laws of thermodynamics—it is only with capitalism that a systematic compulsion to accumulate comes into being that propels an ever-increasing utilization of nature to increase productivity.

This chapter has also argued that Marx's use of transhistorical categories in no way dilutes the historical specificity on his analysis in *Capital*. While concepts such as value can reasonably be seen as internal to capitalism, it is crucial to his project of a critique of political economy that other concepts—such as use-value and concrete labor—are not. In a nutshell, my argument is that the contrast between the two kinds of concepts serves to highlight the contradiction between the capitalist totality and its outside. Therefore, there are good reasons for a critical materialism not to ban the transhistorical use of concepts such as use-value.

Finally, I have pointed to the limits of a pure value-form analysis when it comes to understanding the concrete contexts in which experiences of non-identity or contradiction usually occur. To make sense of these experiences there is a need to supplement the value-form analysis with an analysis of how the capitalist logic becomes embedded in various life contexts, not least the experience of nature.

6

Constellations and Natural Science

What is the relation between the critical theory of nature and natural science? Many environmental threats become visible thanks only to science. The dancing cats in Minamata, for instance, were a riddle until scientists at Kumamoto University were able to connect the dots between the Chisso Corporation, the chemical transformation of mercury by bacteria in the sea, and the nutritional chains that brought this toxic substance to the cats. On the one hand, natural science is of obvious importance in detecting and analyzing environmental problems, but on the other hand science and modern technology have themselves been accused of complicity in causing these problems. To what degree can science and technology be insulated from social critique? Is it unrealistic to envision a new science and a new technology in the future, as Marx and Marcuse did?[1]

These questions clearly have wide-ranging implications for many debates about the environment. Yet critical theory has often been criticized for its failure to come to terms with natural science. The eco-Marxist John Bellamy Foster has blamed this failure on the Frankfurt School's "culturalism," which he sees as rooted in a sociohistorical conception of dialectics that cannot be applied to nature. The Frankfurt School "developed an 'ecological' critique which was almost entirely culturalist in form, lacking any knowledge of ecological science" (Foster 2000: 245). Its restriction of dialectics to the sociohistorical domain, he claims, not only weakened its capacity to intervene in the debate on environmental destruction but also led it to hand over the study of nature to positivism (Foster 2000: vii). A similar critique has been put forward from a social constructivist perspective by Steven Vogel, who argues that the Frankfurt School found itself wavering in regard to natural science—sometimes "accepting (natural) science as unexceptionable when applied to the sphere of nature" but sometimes "wanting to go further and criticize natural science as such" (Vogel 1996: 3). While criticizing natural science as an accomplice to the domination of nature, it failed to explain how a knowledge of nature could be obtained that might substitute for natural science. This failure, he claims, led it to oscillate between treating nature as an ineffable other and as a social construct.[2]

Although it is true that the classical Frankfurt School critical theorists hardly engaged with natural science, these criticisms are wrong in suggesting that theoretical reasons prevented them from doing so. In this chapter, I will argue that the Frankfurt School possesses sufficient theoretical resources for coming to terms with natural science in a satisfactory way. I start by going back to Lukács, since it is in his writings that the problem of how to relate dialectically to nature without the aid of Engels's dialectics of nature—the so-called "Lukács problem" (Foster et al. 2010: 224)—comes to the fore most clearly.[3] I then turn to Adorno's idea of constellations and show how it can be developed into a model for relating to natural science that is preferable to alternative attempts to bring natural science into Marxism, whether relying on objectivistic conceptions of dialectics, constructivism, dualism, or ideas of a nature-subject. I end by suggesting that constellations also point to a way for critical materialism to solve the general problem of the outside.

The Lukács Problem

While Lukács and other Western Marxists generally rejected the causal materialism associated with Engels, they largely left the question how dialectics should relate to nature unanswered. In his classic 1923 work *History and Class Consciousness*, Lukács promotes a subject-centered notion of dialectics as a tool for criticizing reification. Dialectics alone, he argues, has the potential of dissolving the reified thing-forms imposed on the world by bourgeois thinking. It does so by relating them to the totality of the historical situation as it appears to the subject of revolutionary praxis, showing that they are mediated through this totality and determined by it. Since this dialectics is centered on the subject it is only applicable to nature to the extent that the latter becomes involved in human action (Lukács 1971: 24).

The Lukács problem arises because a general critique of reification seems impossible to combine with a restriction of the dereifying tool, dialectics, to society. While Lukács's conception of dialectics requires a restriction of dialectics to human praxis, the program of criticizing reification seems to imply that the critique should be extended to natural science, the prototype of reifying bourgeois thinking. Connected to this methodological dilemma is the question of the ontological status of nature. Is nature per se constituted in such a way that it can be satisfactorily grasped in a contemplative mode? Is it, so to speak, naturally made up of things that cannot be dereified?

Discussing how serious this problem is in *History and Class Consciousness* is complicated by passages that modify or qualify the restriction of dialectics to human praxis. Firstly, Lukács's stress on the subject-object relation did not prevent him from inserting an isolated passage that in passing acknowledges the possibility of an objective dialectic operating in nature independently of humans (Lukács 1971: 207).[4] Secondly, his repeated statement that nature is a "social category" (e.g., Lukács 1971: 234) opens a quite different door to dialecticizing nature since it suggests that our knowledge of nature, if not nature itself, is subjected to the same dialectical processes that shape society. While not irreconcilable, the two statements pull in opposite directions.[5] The former acknowledges an objectivistic dialectics, but the latter points to knowledge as a social construction.

Spotting the weakness of Lukács's position, critics like Abram Deborin, Laszlo Rudas, and others accused him of lapsing into dualism by separating society from nature. Lukács penned an angry response known in English as *A Defence of* History and Class Consciousness: *Tailism and the Dialectic* (below the *Defence*) sometime in 1925 or 1926. The manuscript was left unfinished as he chose to kowtow to party orthodoxy and eventually repudiated *History and Class Consciousness*. Apart from the *Defence*, few texts exist where he even mentions the latter work. The most notable exception is the preface he wrote for the book in 1967, which, however, is little but an extensively argued rejection of his own early work, a text where he goes to great lengths to castigate himself for a variety of errors, including having viewed "Marxism exclusively as a theory of society, as social philosophy, and hence to ignore or repudiate it as a theory of nature" (Lukács 1971: xvi). In contrast to this preface, the *Defence* is truly a *defense* of *History and Class Consciousness*.

The *Defence* presents several arguments related to nature. Two appear to me to be particularly important. Firstly, Lukács clarifies that when he had called nature a social category in *History and Class Consciousness* he wrote "only of *knowledge of nature* and not nature itself" (Lukács 2000: 97). That nature is a social category simply means that there is no socially unmediated relationship of humans to nature, not that nature lacks objective, independent existence. Whatever framework we use for studying nature—for example, an objectivistic dialectics of nature or sciences of the positivistic type—will therefore always be embedded in a social dialectics. This first argument, then, forcefully asserts the necessity of a subject-centered dialectics regarding our knowledge of nature. He therefore defends his decision in *History and Class Consciousness* to characterize the interaction of subject and object as the decisive dialectical categories rather than Engels's so-called laws (Lukács 2000: 102–12).[6]

But how should such a dialectics relate to natural science? To tackle this question, he introduces a second argument. Natural science—like all products of consciousness—is determined by society but the course of its transformation cannot be known in advance (Lukács 2000: 113–18). The non-dialectical traits of existing natural science may have to be abandoned if we come to the realization that nature is more changing and less ahistorical than we thought, but he leaves it open whether this will ever happen. Even the socialism taking form in Russia must therefore use natural science in its bourgeois form for the time being. To dismiss it as a merely bourgeois form of thinking and jump immediately to another, socialist science would be an illegitimate shortcut and "false relativism" (Lukács 2000: 113f). Here Lukács seems to repeat Marx's argument in the Paris manuscripts that a unitary science is a utopian project not achievable here and now.[7]

Taken together these arguments amount to an important clarification of the position in *History and Class Consciousness*. While acknowledging that our knowledge of nature is socially mediated, Lukács argues that for the time being the existing methods of natural science are legitimate and that we therefore must accept a dualism of methods for nature and society. At the same time, he holds out the prospect of a different natural science in the future, one that will be self-aware of its own social determination and realize the unitary science that Marx envisioned.

Why are the natural sciences so hard to transcend compared to, say, the social sciences? Andrew Feenberg (1999, 2014: viiif, 135ff) has suggested an answer to this question that defends what he sees as Lukács's dualism on methodological grounds. Lukács's distinction between nature and history, he argues, is a practical response to the fact that society alone is a realm that we can transform by becoming self-conscious. When societies become conscious of the continency and irrationality of the institutions that govern them, they can try to change them.

> It is because, in the social domain, we *are* in the strongest sense the object that knowledge of society is self-knowledge and as such transformative. In this domain becoming self-aware immediately alters the logic of collective action. No comparable change in natural scientific law results from dereifying self-knowledge in that sphere.
>
> (Feenberg 1999: 88)

To Lukács, the problem is therefore "not with scientific reason *per se*, but with its application beyond the bounds of its appropriate object, nature" (Feenberg 1999: 62). While our understanding of, say, gravity or photosynthesis wouldn't change the natural laws, things are different with society, where our self-awareness directly affects society in a way that is not true of nature.

Feenberg stresses that Lukács's dualism is methodologically motivated rather than ontological: "He is not telling us what kind of *being* nature is but on what terms it can be *changed*" (Feenberg 2014: 135). The conclusion is nevertheless that nature is "*essentially* reified and knowledge of it is destined to remain permanently 'contemplative'" (Feenberg 2014: 137). While reified social forms can be transformed once we understand them, nature is not modified by consciousness. This leads to Feenberg's original suggestion that reified thought should be given a recognized place in dialectical thought, as part of the latter: "Reification is ... not the 'opposite' of dialectics, but a moment in it" (Feenberg 2014: 116). Hence, we should not picture socialism as the overcoming of reification, as Lukács does, but as its complex reembedment. "Socialism is a reorganization of the society around a dialectical mediation of the reified capitalist inheritance ... reification is never completely eliminated" (Feenberg 2014: 134). In practice, this embedment means opening up our engagement with nature to broader public participation and critique (Feenberg 2014: 214).

Feenberg's argument is strong and his suggestion that reified bits of knowledge should be incorporated as moments of the dialectic helpfully points forward to the idea of constellations that include natural science, which I will discuss below. Despite this it is hard to regard his solution as the last word concerning the Lukács problem. To begin with, it simply defends the dualism, leaving the practical question unanswered of how a dialectics based in society should relate to the natural sciences in actual research, where it is often hard to make a neat separation between the realms of nature and society. Theoretically, it abandons the goal of dereifying nature, which I have already stated that a critical theory of nature cannot do. My most important consideration, however, is that Feenberg, like Lukács, takes his point of departure in practical materialism. The reason that they view dialectics as more suited to society than nature is that they see the task of dialectics as giving us a conceptual grasp of the movement of change generated by people's self-consciousness. This view of dialectics is not necessarily persuasive to causal and critical materialists. A causal materialist would hardly find the argument about self-awareness decisive and in critical materialism the task of dialectics is not to conceptualize historical change but to undermine reified conceptual constructions.[8]

However, defending a (qualified) dualism between nature and society, as Lukács and Feenberg do, by no means exhausts the list of possible ways to respond to the problem. At least three alternative solutions have been proposed—Ernst Bloch's, Vogel's, and Foster's.[9] Bloch's solution is startlingly different from Lukács's. Instead of viewing nature as an object for instrumental or technical activity, he draws on Paracelsus and Schelling to envision nature as an incipient subject with which we should cooperate in creating utopia (Bloch 1995: 669–74, 686–9). Thereby he can keep the premise that the dialectic must involve a relation between subject and object and still maintain that nature, indeed cosmos in its entirety, is suffused by dialectics. This solution, however, comes at a high price. As Alfred Schmidt points out, the idea of such a subject working itself out independently of human activity amounts to a "pantheistic-hylozoic conception of a 'nature-Subject', and hence of course to the abandonment of the materialist position" (Schmidt 2014: 59).[10]

Instead of extending the subject-category to non-human nature, an alternative solution is to view the latter as an entirely social construction. This is the position proposed by Vogel (1996) who argues that a solution should take its point of departure in Lukács's recognition that nature is a social category. What is needed is to extend the ideology critique developed in Western Marxism to the natural sciences by exposing the role of social subjects in constituting nature. According to Vogel, critical theory has been reluctant to turn to social constructivism since it appears to contradict materialism, but it is necessary if it wants to remain true to its ambition to counteract the reification of nature. Doubts, however, can be raised about the viability of Vogel's solution. Not only does it run into the objection offered by Feenberg that nature will always appear objective since it remains untouched by self-consciousness, it also risks trivializing the severity of ecological crises. As Andreas Malm (2018) argues in his critique of Vogel, acknowledging the objectivity of nature and natural processes is crucial to understanding environmental problems. However, rather than relying on science in realist fashion, as many critics of social constructivism do, my objection against constructivism is that it misses the primacy of the object, the intransigence of matter in relation to thought which means that no identity can be established between them. This is missed in Engels's dialectics of nature no less than in Vogel's social constructivism, although one veers too much to the side of nature and the other too much to society.[11]

Still another way to link together dialectics and nature is by reviving an objectivistic dialectics of nature along Engels's lines but in a less deterministic form, as proposed by Foster. We will have a closer look at this solution in the next chapter, but already here we can state that it shares some of the unsatisfactory traits of other objectivistic forms of dialectics, especially in regard to the question of what role to assign to the subject and to praxis. As I show in the next chapter, rather than a solution in its own right, it amounts to a compromise between practical and causal materialism that to a large extent leaves the weaknesses of both unremedied.

None of the solutions we have looked at so far is free from problems. A satisfactory solution, I suggest, must preserve the possibility of a dereifying critique regarding nature. The solution can therefore neither be to rein in dialectics to society, nor to follow Engels in attempting to discover an objectivistic dialectics in nature. Both options imply acquiescence to reifying nature. On the other hand, if nature is what cannot be reduced to the subject, then social constructivism and "pantheism" are also ruled out. Below I will argue that the best way forward—and the only one that is consonant with a critical materialism centered on the primacy of the object—is to build on a usage of concepts that Adorno referred to as constellations.

Constellations

When critical theory inquires into nature, it tries to avoid reifying it and making it available for technical manipulation and marketization. Critical theory therefore needs a language that prevents its objects from becoming useful to dominant institutions such as capitalism or the state. But how can concepts break free of the identity-thinking that, as Adorno writes, exists as an impulse or tendency in all thinking? Resorting to feeling and intuition in romantic fashion will not do, since romanticized nature fits all too well into the system. Rather than advocating a return to immediate, pre-conceptual experience, concepts must be used, even as they are criticized.

A model for how this is possible is provided in Adorno's discussion of constellations (Adorno 1975: 62, 164–8). The idea of constellations was inspired by Benjamin, who in *The Origin of the German Tragic Drama* had written that he wanted to present his ideas in the form of configurations. Ideas, he suggested, were not like abstract concepts that simply stated the common denominator of the phenomena that they sought to cover. Instead, they were made up of the phenomena themselves, just like constellations were made up of stars (Benjamin 1985: 34). Rather than defining an idea abstractly, he therefore sought to present it by going from one individual phenomenon to the other with a minimum of abstract bridging concepts. Ideas, he claims, "come to life only when extremes are assembled around them" (Benjamin 1985: 35), when approached through instances that retain their uniqueness. In the Arcades project, this procedure took the form of what he—following the surrealists—called the montage, a technique that juxtaposes heterogeneous elements in order to disrupt and ruin the mythic semblance of self-sufficiency and closure in each. The book he aimed at would take the form of a constellation of themes—such as the *flaneur*, fashion, and exhibitions—that would mutually relativize each other and, in their encounter, bring about an awakening from the mythic dreamworlds in which they were embedded.[12]

Adorno's usage of constellations is indebted to Benjamin's notions of configurations and montage, but he reorients them to a model for how concepts might approach objects in a non-oppressive way. A constellation, to him, is a way of using concepts that doesn't subsume the object, but, in line with the latter's primacy, encircles it by a plurality of conceptual elements that need not be logically connected to each other. "As a constellation, theoretical thought circles the concept it would like to unseal, hoping that it may fly open like the lock of a well-guarded safe-deposit box: in response, not to a single key or single number, but to a combination of numbers" (Adorno 1973a: 163). In the rationale for constellations, we rediscover the reason for Adorno's use of the *chiasmus* as a basic stylistic move: it is a constellation in miniature, on the level of the sentence. Like the *chiasmus*, the constellation juxtaposes incongruous perspectives, demonstrating that they are essential to grasping the different sides of objects (Buck-Morss 1977: 57, Pensky 2005).

Adorno's constellations contrast with identity-thinking in at least four ways. The first is that they prioritize the capacity to illuminate the object over the internal logical structure of the system. Rather than logical consistency, what holds a constellation together is its ability to illuminate the contradictory character of its object. "As thinking, dialectical logic respects that which is to be thought – the object – even where the object does not heed the rules of thinking" (Adorno 1973a: 141). Connected with this is a second difference: constellations break with the step-by-step hierarchical progression to more general concepts typical of classificatory thought. Each concept directly relates to the object, illuminating an aspect missed by other concepts. Thirdly, since constellations are open to experience, they cannot be static or fixed. "Truth," as Adorno points out, "is a constantly evolving constellation" (Adorno 2005: 131).

Fourthly, and importantly, constellations may be objective in the sense of being true to objects, but they are not impartial. Springing from experience, the conceptual elements that they mobilize must do justice to the pain that belongs to that experience but which would be rendered invisible if viewed only from the standpoint of the system. A model for how constellations can be deployed in this way can be found in Marx. For instance, the examples of the exploitation of child-labor in the chapter on the working-day in *Capital* build on testimonies, parliamentary reports, and news articles that make us see the logic of capital accumulation in a new light. These elements form a constellation. The critical effect arises from the fact that many facets of the exploitation would have remained invisible if we had focused one-sidedly on the economic categories alone.

Constellations are not an anomaly in Adorno's otherwise negative or critical thinking, despite their ability to render experiences in a positive guise. They are part and parcel of negative dialectics. The differences between constellations and identity-thinking listed above follow from his insistence on the experience of non-identity between the categories of identity-thinking and their outside. The movement between the heterogeneous elements in a constellation may be destructive from the standpoint of the criticized system but has the positive aim of doing justice to the experience of the object in a way that doesn't relinquish conceptual thinking.

A comparison to Hegel can be instructive in helping us see the balance between the destructive and constructive sides of constellations. The template for Adorno's critique of identity-thinking is Hegel's critique of abstract thought. What Hegel calls sublation is the movement of thought from the simple and abstract to the complex and concrete (see Fine 2001: 33). This movement also describes how constellations operate, namely, through concepts that negate each other and thereby illuminate different sides of the totality under consideration.[13] The difference to Hegel is that Adorno's constellations never bring the concepts together in a logically integrated totality. Their splintered state reflects the contradictory nature of reality. Whereas Hegel believes that objects can be fully captured through conceptual mediations, constellations gain their raison d'être from the non-identity between concept and object that prevents thought from stabilizing itself in the form of a system. While Hegel presents dialectics from the standpoint of an already achieved reconciliation, constellations bring out that the world is still unreconciled.

The usefulness of constellations for environmental thinking should be obvious. As surely any instance of environmental destruction shows—from Minamata to contemporary climate change—we are driven to use constellations whenever we try to account for these catastrophes concretely, in a way that summarizes what we need to know about them in order to take a stance and possibly act in relation to them. Scientific reports, charts, models, and concepts have to be mobilized along with glimpses of utopian imagination, history, testimonies, personal memories, or journalistic reportage. We would likely feel that an account that *failed* to refer to such a wide variety of elements would be one-sided.[14]

While no single vantage point captures the object entirely, it would also be a mistake to believe that a constellation in its entirety succeeds in capturing all sides of the object. Not even constellations can achieve full reconciliation with the object. They build, however, on a recognition of misrecognition. Unlike in identity-thinking, the non-identical is highlighted rather than forgotten or covered up. No matter how many perspectives I juxtapose in order to bring out the truth of, say, the pollution of Minamata Bay, I can never pretend that my knowledge is sufficient to grasp the suffering of a single fisherman or dancing cat. The persistence of non-identity is never forgotten.

Constellations and Natural Science

Against the criticism that the Frankfurt School's conception of dialectics forced it to pull back from the study of nature, it should be pointed out that Adorno was far from hostile to natural science. As he himself writes:

> Surely not the least of the tasks incumbent upon philosophy is to help spirit appropriate the experiences of the natural sciences without recourse to amateurish analogies and syntheses. An unproductive gulf exists between the natural sciences and the so-called realm of spirit.
>
> (Adorno 2005: 14)

He also clarifies why natural science should be embraced:

> One argument for the primacy of the object is indeed incompatible with Kant's doctrine of constitution: that in the modern natural sciences *ratio* peers over the wall it itself erects, that it snatches a snippet of what does not agree with its own ingrained categories. Such an expansion of *ratio* unsettles subjectivism.
>
> (Adorno 2005: 251)

Natural science helps us criticize our categories and break out of idealism. There is thus hardly any fundamental theoretical incompatibility between negative dialectics and natural science. Unlike what critics claim, the

Frankfurt School is not per se hostile to natural science. What it *is* hostile to is the mutual compartmentalization of thought and experience that lets each remain in abstract isolation from each other. This is what Horkheimer (2002b: 183) argues when he states that natural science has a legitimate place within a dialectical consideration of totality that "weaves" together different disciplines.

Although Adorno seldom refers to natural science, his idea of constellations indicates a fruitful way for critical theory to relate to such science. Nothing prevents bits and chunks of scientific knowledge from being included in a constellation. Thinking dialectically isn't to do the work of natural science but to *insert* these bits and chunks into the constellation in a useful and illuminating way, without letting them take on the status of absolutes. Such a way of thinking about natural science is reminiscent of how Feenberg theorizes the place of reified thought as "not the 'opposite' of dialectics, but a moment in it" (2014: 116). This, I suggest, is what Adorno may have meant in the quote above when he wrote that it is a task of philosophy to "help spirit appropriate the experiences of the natural sciences" (2005: 14). This way of thinking by means of constellations is not only preferable to the rigid, dualistic separation between nature and society, with each realm possessing its own proper method. It also shows clearly why the rejection of an objectivistic dialectics of nature doesn't have to mean relinquishing dialectical thinking about nature.

How does this differ from how causal materialism and practical materialism relate to natural science? Let me start with causal materialism. Whereas an objectivistic dialectics of nature would have to challenge existing forms of natural science on their own turf by imitating and competing with them, constellations offer a different model for how to think nature dialectically. Rather than imitating natural science by searching for impersonal laws operative in nature or society, they become critical tools that illuminate contradictions in our experiences related to nature. Such tools are indispensable for a critical theory that seeks to be practical and emancipatory.

This is not to say that critical theory cannot challenge science. But it does so by juxtaposing science with parts of its extra-scientific context rather than by competing with it. This has implications for the neutrality of science and technology. Science ordinarily achieves its semblance of neutrality by bracketing the wider context in which it is conducted. This context is unbracketed in part when science is inserted into a constellation. In contrast to science, constellations do not aspire to objectivity in the sense of neutrality or impartiality but to be true to the contradictions that we sense in the object. Constellations become necessary as soon as we must take a stance on any concrete issue, because they alone promise us a view of the matter that is not one-sided. Science, by contrast, can provide no guidance on its own to such a stance as long as it retains its abstract form, that is, its isolation from social context. Nuclear power, for instance, is hard to understand solely on the basis of science, without concern for its use in, say, war. Even when science no longer pretends to be pure but is actively enlisted in support for normative or political goals, it can be challenged by constellations that illuminate the contradictions that these categories give rise to.

Constellations and Dereification

The above remarks have clarified the contrast between constellations and objectivistic conceptions of dialectics in causal materialism. But how do constellations differ from how dialectics is deployed in practical materialism? Looking into this is relevant, since it was in practical materialism that the Lukács problem appeared with full force, as a result of the rejection of Engels's causal materialism. We have seen how Lukács himself seemed to land in a dualist position in which he, at least for the time being, had to abandon the goal of dereifying nature and accept positivism as valid in natural science.

Constellations avoid that dualism. Their elements mutually relativize each other regardless of whether they are natural or social. Importantly, that means that constellations can help dereify nature. Even if constellations include reified bits of knowledge as elements, the overall effect is dereifying since no element asserts itself as absolute. There is therefore no reason to restrict a dialectical, dereifying critique to human affairs. If nature too can be reified, then the goal of liberation from reification requires dialectics to be applied to nature as well.

Reified objects are viewed through the lens of the identity-thinking ingrained in dominant categories. In most cases this means that they are severed from their history as well as their sensuous qualities. Constellations counteract reification by reminding us of what the thing-form invisibilizes and thereby become a critique of that thing-form.

What prevented Lukács from taking this route toward dereifying nature? I can think of two reasons. One obstacle may have been his tendency to see reification narrowly as the imposition of thinglikeness on *human* creations, a tendency reflected in the fact that he often appears to view second nature as artificial and first nature as genuine. As against that position I have argued that nature too can be reified. Lukács's restrictive view of reification also colors Feenberg's account. To him dereification is linked to a movement of self-consciousness that is lacking in nature. Although he points in the direction of constellations when he argues that reification should be retained as a moment within dialectics, his use of the term "reification" differs from mine. To him the thinglikeness that science imposes on nature is unproblematic since reification is primarily a matter of mistaking social relations for things (Feenberg 2015: 490). From my perspective, however, bringing in bits of science within a constellation is exactly what at the same time *dereifies* them. Rather than linking dereification to self-consciousness, I see it as the restoration to things of aspects that are excluded from view by dominant categories, regardless of whether those things are natural or social.

An additional reason for Lukács's unwillingness to take this route may have been his practical materialism, which made him see dialectics as a tool for weaving together theory and praxis into a dialectical unity or totality. It is from that perspective that the irreducible otherness or objectivity of nature presents itself as a stumbling block, as an alien element that refuses incorporation in the meaningfulness of totality. But this obstacle disappears in critical materialism, which instead uses experiences of non-identity, expressed in constellations, to undermine totalities. The crucial move is to move from a conception of dialectics stressing unity to one geared to highlighting non-identity. This move requires a shift from a materialism in which praxis is designated as the material element and to which nature appears as alien, to another materialism based precisely on this alien element and its capacity to subvert dominant categories.

To illustrate how constellations can contribute to dereification—and also to hint at the difficulty of setting up constellations in a way that adequately captures contradictions—we might mention Anna Tsing's study *The Mushroom at the End of the World* (2015), which follows the *matsutake* mushroom through its ecological entanglements with industrial forests and global capitalist supply-chains from the salvage economy of former refugee mushroom pickers in Oregon, Yunnan, and Finland to the dinner tables in Japan. In my view it does a remarkable job in evoking the larger whole of global capitalism through the juxtaposition of meticulously studied local contexts. She herself refers to the book as a book in assemblage form, but her procedure of juxtapositions can also be compared to what Adorno called constellations. Like constellations, her assemblage to a large extent succeeds in dereifying its object, the mushroom, which is presented to the reader in all its pungent odor and with a multiplicity of aspects that a mere focus on the commodity form would have obscured. At the same time, there are limits to the dereification that her study effects, since contradictions are largely absent from it. A reader may be forgiven for walking away from it with a sense of wonder, not only at the marvelous complexity of ecological entanglements but also at the marvel of capitalism. Constellations, as we recall, are meant to be true to the pain that arises from the imposition of dominant forms on reality but in her work little of that pain can be felt. The critique of capitalism that isn't openly voiced is instead expressed indirectly, in the book's fascination with the post-apocalyptic imagery evoked by the salvage economy of mushroom picking, which is described as a way of surviving in ruins. Clearly, there are anxieties under the text's surface.

As I've argued above, constellations steer clear of the weaknesses of both objectivistic conceptions of dialectics and the dualism toward which Lukács and Feenberg tend. They also steer clear of the main drawbacks of other proposed solutions to the Lukács problem such as Bloch's idea of a nature-subject or Vogel's social constructivism. Firstly, constellations remove the need to posit nature as a subject writ large. Rejecting this solution to the Lukács problem doesn't mean that I reject recognizing subjecthood in nature. What I reject, however, is the idea that we must presuppose such subjecthood in order to apply dialectics to nature. Secondly, the fact that constellations always relativize our conceptual grasp of nature does not mean that nature is a mere social construct. Social constructivism misses the primacy of the object, the fact that nature is characterized by an objectivity that cannot be fully reduced to the way it is socially perceived and defined. Here a critical theory of nature sides with Lukács: what is socially constructed is our knowledge of nature, not nature itself. When Vogel criticizes Adorno for having to treat nature as an ineffable other by relinquishing all foundations for a knowledge of nature, he misses the important role that constellations play in Adorno's thought, a role that is precisely to use concepts to help us think dialectically about experience.

Thinking the Outside

The idea of constellations not only shows how dialectics might relate fruitfully to natural science, but also brings us closer to coming to terms with the more general problem of how to think the outside. It is a misconception that negative dialectics must restrict itself to a mere critique. Instead it confronts the categories of the criticized system with the objects they claim to represent. It goes hand in hand with a reflection on experience that brings concepts into felt contact with the object and tries to do it justice. Far from being merely negative, such reflection sensitizes thought to its environment and helps it grasp it concretely.

The answer to how can we know what lies outside our conceptual systems is that we ourselves *already* belong to this outside through our experience. Since immanent critique takes the confrontation of concept and object as its point of departure, it relates to the outside from the start. Strictly speaking, the problem of how to gain access to the outside never arises, since such access is the source of the pain that is the very premise of the critique. It is identity-thinking rather than negative dialectics that has problems thinking the outside. The impression that negative dialectics is unable to think the outside presupposes that such thinking must take the form of identity-thinking, but it is precisely that presupposition that negative dialectics rejects. Instead, its answer to how to think the outside is provided by the idea of constellations.

As I argued in Chapter 2, Marxism offers four options for theorizing nature as an outside of capitalism: extending the totality to at least parts of nature through a logic of positing presuppositions, supplementing the analysis of capital in eclectic fashion with other theories treating nature, varying the levels of abstraction, or constructing constellations that express the non-identity between concept and object. All four options help us talk about the outside of capitalism, but in my view the fourth is of key importance. It is from that point of view that we can see clearer *why* the others are useful.

Firstly, the logic of positing presuppositions is excellent for highlighting why nature and other seemingly non-economic domains are functional for capitalism. This logic, however, does not in itself solve the problem of how to think the outside since it merely expands the boundaries of the capitalist totality. Combining it with constellations would enable it to approach the object not only from the point of view of its functionality for the system but also from the point of view of its non-identity with it. Juxtaposing perspectives in this manner is precisely what constellations are good at. To some extent, Marx and Marxists following him made use of precisely such a combination when they pointed to the pain, injustice and violence that is hidden behind the façade of capitalism—the instrumentalization of nature, the brutality of primitive accumulation, the toil of reproduction, and so on. The functionality of nature and other seemingly non-economic domains for capitalism was highlighted not only for analytic purposes, but also to criticize and indict the system that is dependent on exploiting it.

Secondly, constellations help justify eclecticism to the extent that the theories that they bring together can be shown to be necessary to illuminate contradictions in the object. Constellations are already from the start eclectic, capable of conjoining conceptual elements in a way that ensures the overall character of critical rather than traditional theory. While this opens the door for using natural science in critical theory, the purpose is not to contribute to a universal theory of nature and society by fitting together the fragments into a new system. Instead, constellations sharpen our awareness of non-identity precisely by letting their elements contradict and mutually relativize each other. The elements are not brought together as pieces of a jigsaw puzzle that must be fit together but are meant to clash.

Thirdly, shifting levels of abstraction can be useful, but to have critical effect this thought operation needs to relate to the negative totality made up of the predominant system of real abstractions, since it is primarily in relation to that system that objects are experienced as non-identical. It is true that Marx often employed shifting levels of abstraction, but as we saw in the previous chapter he did so to criticize the present rather than to construct transhistorical or universal theories. The point of referring to transhistorical concepts like material wealth was not to abstract from capitalism but rather to bring out the contradiction between them and the rule of value in capitalism. In other words, a transhistorical concept on a high level of abstraction was brought in to highlight the non-identity between itself and value. The clash here, I suggest, is similar to the clash between conceptual elements in a constellation.

Constellations offer a way of thinking about nature and other parts of capitalism's seemingly non-economic outside that preserves the critical potential in the experience of non-identity. At the same time, they preserve the strong points of the other three options, allowing for insights into the systemic interconnections regulating the capitalist economy and its relations to its outside as well as for bringing in additional theories where needed.

Conclusion

This chapter started with the Lukács problem: how can natural science be accommodated within a dialectical approach, if such science is the very prototype of reifying bourgeois thinking? I have argued that constellations offer a way to come to terms with this problem since they are applicable wherever a relation of non-identity exists between concepts and objects, regardless of whether those objects are classified as social or natural. Nothing says that they must confine themselves to the social realm. They can be applied to *any* object, whether social or natural, and natural science can very well form part of them. Rejecting the objectivistic dialectics associated with Engels therefore does not mean handing over the study of nature to positivism, as Foster claims. Engels's mistake was not to apply dialectics to nature, but to conceive of dialectics as a set of laws objectively active in society as well as nature rather than as a critical procedure obtaining in the relation between concept and object. Constellations are dialectical since their elements mutually mediate and relativize each other, rather than existing as self-sufficient or reified facts. Bringing them together and letting them highlight each other's shortcomings avoids reifying them.

If I am right that constellations help us solve the Lukács problem, there is no need for a Marxist analysis to shun the subject of nature, or to reduce it to a mere social construct or reify it by identifying it with natural laws. As mentioned, Vogel claims that critical theory due to its critique of natural science and of Engels's objectivistic dialectics of nature had to vacillate between viewing nature in Hegelian fashion as a social construct and romantically as an ineffable other. Horkheimer and Adorno, he claims, failed to explain how a knowledge of nature could be secured that might substitute for natural science, and Adorno's subsequent replacement of nature by non-identity aggravated the problem since the latter per definition cannot be known (Vogel 1996: 3f). This claim is incorrect, since one of Adorno's key aims is precisely to show modes of knowledge beyond science are possible that pay respect to the primacy of the object. He never claims that the non-identical cannot be known, and it is vastly off the mark to portray him as a romantic who reduces nature to a static otherness untouched by historical mediation. By approaching the object through constellations, there is no need to leave it unsaid or apprehend it as static.

Constellations also provide an answer to how critical theory can approach the related problem of how to think the non-capitalist outside. In theorizing this outside, Marxism has been torn between, on the one hand, building transhistorical theories in which the ambition to explain history has often overshadowed the critique of capitalism and, on the other, limiting itself to an analysis of capitalism that fails to illuminate the relation to the non-capitalist outside. Adorno's negative dialectics offers a way beyond these two alternatives. Since it takes the confrontation of concept and object as its point of departure, it involves a relation to the outside from the start. While immanent critique focuses on criticizing established systems it is not limited to replicating their self-understanding. As he points out, such critique is also transcendent, not in the sense of relying on external yardsticks, but by virtue of its reference to the object.

7

Eco-Marxism's Return to Marx

With the revival of interest in Marx in relation to ecology in recent decades, the idea that he was a simplistic supporter of industrialism with nothing of interest to say about nature can safely be discarded. In particular, the efforts of eco-Marxists like John Bellamy Foster and Paul Burkett have contributed greatly since the 1990s to defending his status as a pioneering ecological thinker with an abiding interest in natural science. In the process of doing so, they have also made great strides in clarifying the ecologically destructive side of capitalism, highlighting capitalism's relation to nature as the site of its most central contradiction (for this latter claim, see Burkett 2014: 176–82, Saito 2017a: 259).[1]

This chapter focuses on Foster, probably the most influential representative of the present generation of eco-Marxists. His seminal 2000 book *Marx's Ecology* has been enthusiastically received and is already an Eco-Marxist classic. But rather than discussing the admirable aspects of this work, I will focus on an issue where I believe it falls short, namely, its attempt to resurrect the idea of a dialectics of nature. This attempt is accompanied by a barrage of criticism against Lukács, the Frankfurt School, and other Western Marxists, whose rejection of Engels's attempt to extend dialectics to the realm of nature, he alleges, led to a neglect of nature in favor of analyses of culture.

Both Burkett and Foster find fault with the Frankfurt School for its inability to adequately address the present ecological crisis. I have already scrutinized the errors that mar Burkett's criticism of Alfred Schmidt (see Chapter 4). Foster, in a more nuanced argument, aims his criticism at Western Marxism's restriction of dialectics to the realm of society, which he claims represented a turn in idealist direction that abandoned the study of nature to positivism. His remedy is to revive a dialectics of nature of Epicurean inspiration to ensure the possibility of a unity of method in the study of nature and society. That is the argument I want to scrutinize in this chapter.

In this chapter, I argue that Foster's conception of dialectics fails to secure a unity of method and that the critical theory of nature offers better theoretical tools for grasping the relation between nature and capitalism. Rejecting Foster's criticism of critical theory is not to claim that the latter is faultless. I end by suggesting ways in which it may develop to fulfill its potential role as a critical theory of nature, assessing potentially useful contributions of eco-Marxism.

The Criticism of Western Marxism

Marx's Ecology is well known for highlighting Marx's idea of a metabolic rift. Combined with a reading of Marx emphasizing the legacy of Epicurean materialism, Foster uses this idea as the starting point of an ambitious project to bring together natural science and the analysis of capitalism under the aegis of a reconceptualized dialectics of nature. His overall aim is to demonstrate the possibility of a unity of method in the social and natural sciences with the help of Marx's conception of materialism and dialectics. "No other approach," he argues, "has the capacity to integrate a natural-scientific and social-scientific critique that can inform our practice in the Anthropocene" (Foster 2016b). From this vantage point he and his colleagues criticize the Frankfurt School and Western Marxism generally for limiting dialectics to the social-human realm, thereby ceding the study of nature to positivism, ceasing to be materialist in the proper sense and turning in an idealist direction (Foster 2000: 8). The result of their withdrawal of dialectics from nature, he claims, was a "firewall" and "unbridgeable chasm" established between the realms of nature and society that severely limited the ability of Western Marxists to contribute anything useful to the debate on the ongoing environmental destruction (Foster et al. 2010: 216, 242, 485 n18; compare Foster 2013 n45, Foster & Clark 2016a, 2016b, 2016c).

To escape what he sees as the misunderstandings of Western Marxism, Foster goes "back to the foundations of materialism," above all to Epicurus (Foster 2000: viii). His primary aim is to show that dialectics can be fruitfully applied to both nature and society, and that an essential unity of method between natural and social sciences is therefore possible. He points out that Marx's dialectics "was predicated on the ultimate unity between nature and society, constituting a single reality and requiring a single science" (Foster et al. 2010: 215). The chasm between nature and society that arose with Western Marxism was thus "entirely absent in his work" (Foster et al. 2010: 216). According to Foster, Marx's refusal to divorce materialism from natural-physical science:

> establishes what Bhaskar has called "the possibility of naturalism", that is, "the thesis that there is (or can be) an essential unity of method between the natural and the social sciences"—however much these realms may differ. This is important because it leads away from the dualistic division of social science into a "hyper-naturalistic positivism", on the one hand, and an "anti-naturalistic hermeneutics", on the other.
>
> (Foster 2000: 7)

This passage reveals the stakes in Foster's criticism of Western Marxism and shows why unity of method is so important to him. If such a unity cannot be achieved, his criticism would turn back on himself, since—according to his own argument—he would then himself have to cede ground to positivism and risk ending up in an idealist position. To see to what extent Foster redeems his claim that unity of method is possible we need to scrutinize his attempt to give the insights of Epicurean materialism dialectical form.

An Epicurean Dialectics of Nature

According to Foster, the impasse faced by Western Marxism regarding nature can be solved only through a methodological approach that bridges the divide between nature and society. This requires a reconceptualized dialectics applicable in similar fashion to society as well as nature. To understand what he means by such a dialectics, we need to turn to two ideas for which *Marx's Ecology* is famous: that of the metabolic rift and that of the Epicurean legacy in Marx's materialism. While the rift is analyzed in terms of a dialectical relation between capitalist society and nature, Epicurean atomistics, by contrast, suggests an all-embracing, universal dialectics.

Reflecting the tension between these ideas, Foster presents two different portrayals of what the dialectics of nature that he is proposing would look like. The first portrays dialectics as unfolding primarily in the relation *between* humanity and nature. This is consistent with his emphasis on the metabolic rift and with Marx's emphasis that dialectics centrally involves human praxis, taken in a wide sense as including the way labor as a metabolism with nature is organized. A problem for Foster is that this sort of dialectics wouldn't in principle be different from the praxis-centered dialectics associated with Western Marxism. Dialectics would still be driven above all by human praxis, its applicability to large parts of nature being a by-product of the increasingly wide scope of humanity's metabolism with nature.

Secondly, Foster also suggests the possibility of working out a dialectics *of* nature as such, showing how nature develops dialectically even without human interference. Foster admits that Marx himself tended to link dialectics primarily to human praxis but argues that Marx nevertheless acknowledged the possibility of dialectics operating ontologically in nature itself (Foster 2000: 2, 114). Engels's materialist dialectic was therefore legitimate when viewed from the standpoint of his and Marx's basic philosophical outlook—although Foster qualifies this assessment by adding that Engels overemphasized the deterministic and mechanistic aspects of the dialectics of nature (Foster 2000: 230). To reconstruct a more open-ended dialectics, Foster turns to Epicurus, who was the subject of Marx's doctoral thesis along with Democritus and who Foster argues was a more decisive influence on Marx than the deterministic/mechanistic materialists of the Enlightenment period, such as d'Holbach, Diderot, and de la Mettrie.

Epicurus's atomistics helps Foster avoid both a restriction of dialectics to the human realm on the one hand and a deterministic conception of dialectics on the other. Unlike Democritus and the Enlightenment materialists, Epicurus opposed determinism in nature by allowing for an unpredictable "swerve" or deviation among atoms (referred to by Lucretius as *clinamen*). Far from being governed by iron laws, Epicurean materialism is therefore characterized by open-endedness, contingency, and unpredictability. Against Western Marxism, which rejected the idea of a dialectics of nature because of its determinism, Foster argues with reference to Epicurus that it's possible to be a materialist without being a determinist (Foster 2000: 2, 33–65). In trying to work out a non-deterministic materialism, he is part of a broader trend today that stresses contingency and in which we also find Althusser and contemporary "new materialism."

Foster argues that the Epicurean influence on Marx was overlooked by later Marxists. Neglecting the Epicurean legacy resulted in interpretations of Marx's materialism that were much more rigorously deterministic than Marx himself had envisioned. Rather than Epicurus, it was Enlightenment materialism that inspired Georgi Plekhanov's influential interpretation and led to the positivistic, mechanistic character of official Soviet-style Marxism, which increasingly came to hold sway in the 1920s. Against this Marxism, and also against Western Marxism—which in Foster's view overreacted to this mechanistic materialism by jettisoning materialism *tout court* and turning in an idealist direction—Foster resumes what he sees as Marx's original project of developing a dialectic inspired by Epicurus and contemporaries like Darwin. This implies a clear rejection of the view that dialectics exists only in humankind's relation to nature, not in nature itself, and a partial rehabilitation of Engels's dialectics of nature.

Natural Praxis

But how does Foster bring together the Epicurean and the praxis-centered models for a dialectics of nature? To avoid dualism and demonstrate the possibility of a unity of method, he needs to show that they converge in a similar notion of dialectics. This means that the question how Epicurean materialism can be given dialectical form becomes crucial.

What, then, is required to make materialism dialectical? In general, Foster's argument is that dialectics is our tool for grasping a changing environment. But exactly what, in the various ways people have tried to understand change, is it that makes understanding dialectical? This question is far from clear: while Foster rejects the "mechanistic" view of dialectics as objective laws operating in nature he also rejects the restriction of dialectics to the human subject which he sees as typical for Western Marxism. Foster makes the following distinction between materialism in general and dialectical materialism. The former "sees evolution as an open-ended process of natural history, governed by contingency, but open to rational explanation." The latter, by contrast, "sees this as a process of transmutation of forms in a context of interrelatedness that excludes all absolute distinctions" (Foster 2000: 16). To qualify as dialectical, Epicurean materialism or a materialism derived from it would have to conform to the latter definition. But this distinction is hardly sharp. It also seems insufficient to establish what makes materialism dialectical. Ideas of "transmutation" and of "interrelatedness" are not unique to dialectics. Surely, these or similar ideas are also present in many manifestly *un*-dialectical approaches, such as, say, actor-network theory or the Buddhist-inspired ontology of Kitaro Nishida.

One reason that the definition seems insufficient is that it leaves out the role of praxis and critique. As Foster himself remarks, Marx's own dialectics was not contemplative. Marx's emphasis "was overwhelmingly on the historical development of humanity and its alienated relation to nature, and not on nature's own wider evolution … he tended to deal with nature only to the extent to which it was brought within human history" (Foster 2000: 114). His materialist criticism of Feuerbach's contemplative materialism was, Foster points out, also a criticism of Epicurean materialism (Foster 2000: 112). But if superseding the contemplative attitude through praxis is central to a materialist dialectics, can there be a dialectics of nature in itself, a dialectics that would proceed without human involvement? It is hard to find a clear overall answer in *Marx's Ecology*, but three passages suggest what an answer might look like.

First, Foster suggests in a brief remark that Epicurus's "swerve" turns the universe into a world of freedom and self-determination characterized by "alienated self-consciousness" (Foster 2000: 55). Being a realm of self-consciousness nature might be seen as similar to human history and hence a proper domain for dialectics—a variant of the Blochian idea of nature as a nature-subject. A problem with this idea is that it restricts the scope of dialectics to the parts of nature where self-consciousness can be detected, leaving it unclear how the rest of nature should be dialecticized.

A second argument is that the realms of nature and society are united by what Lucretius refers to as *mors immortalis* (immortal death, i.e., never-ending change).

> If the materialist conception of nature and the materialist conception of history remained integrated in Marx's practical materialism, it was primarily

... through the concept of "mors immortalis" (immortal death), which he drew from Lucretius, and which expressed the idea that, in Marx's words, the only eternal, immutable fact was "the abstraction of movement," that is, "absolutely pure mortality".

(Foster 2000: 114)

What unites the two realms of nature and society is thus said to be change, or, as Engels put it, the fact that "nature does not just *exist*, but *comes into being and passes away*" (Engels 1987b: 324). By itself, however, this idea does not require a break with the contemplative attitude. The idea of *mors immortalis* can easily be criticized as an abstraction that would be eminently graspable from a contemplative standpoint.

Foster's third argument is stronger. Dialectical reasoning, he writes, can "be viewed as a necessary element of our cognition, arising from the emergent, transitory character of reality as we perceive it" (Foster 2000: 232). Here the claim is that the natural and social sciences must use the same heuristic device, dialectics, to grasp the movement of their subject matter. This argument too, however, is unrelated to praxis. Furthermore, it seems to tie dialectics to the human subject, implying that nature can be viewed as dialectical only insofar as it becomes the object of human reasoning. Foster thus offers a patchwork of different arguments that all suggest ways in which dialectics may be extended to nature, but problematically none of the three passages clarifies in what sense the dialectics that he advocates would be connected to praxis.

A possible interpretation would therefore be that Foster doesn't see praxis as a necessary ingredient in dialectics at all, but the door to such an interpretation is slammed shut by Foster's later works, such as the 2010 book *The Ecological Rift*, where he and his co-authors Brett Clark and Richard York advocate a dialectics of nature based on a sensuous "natural praxis" and an ecological perspective spanning nature and society. More explicitly than *Marx's Ecology*, this work rejects contemplative materialism, which, the authors believe

> can be as destructive as idealism. A materialism unrelated to praxis and divorced from dialectical conceptions is ... a mere mechanical myth and can itself be a tool of domination. What is needed is an expansion of our knowledge of the universe of praxis.
>
> (Foster et al. 2010: 247)

This forceful formulation suggests a close affinity to the praxis-centered dialectics Foster elsewhere associates with Western Marxism. The question is how the dialectics advocated by Foster and his co-authors differs from that of their opponents. An answer is provided in their argument that Marx's materialism, unlike that of the Western Marxists, isn't based narrowly on social praxis but on a "natural praxis"—the latter being "a much larger concept of human praxis that encompasses human activity as a whole, that is, the life of the senses" (Foster et al. 2010: 230). Quoting a line by the young Marx—"In hearing nature hears itself, in smelling it smells itself, in seeing it sees itself"—they comment that here "[t]he senses are nature touching, tasting, seeing, hearing, and smelling" (Foster et al. 2010: 227f), suggesting that human self-awareness is but part of a larger self-awareness or subjectivity in nature. This natural praxis, the authors argue, is linked both to the senses and to an expanded ecological perspective emphasizing interaction and connectedness. In an ecological perspective, humans are not separated from nature—"the biosphere is constitutive of our own existence even as we transform it through our actions." Such a perspective is not only dialectical in the sense of stressing the interconnectedness of all living things but also helps breaking free from the strictures of a merely social dialectics—it "constantly seeks to transcend the boundaries between natural and social science." An ecological dialectics, in other words, has the potential to be the unifying "single science" that the young Marx demanded and to vouchsafe the "unity of method" Foster aims at. "The development of ecology as a unifying science is pointing irrefutably to the validity of Marx's original hypothesis that in the end there will only be 'a single science'" (Foster et al. 2010: 246f).

The idea of natural praxis clarifies that Foster and his co-authors think of the dialectics of nature as a subject-object dialectics along the lines stressed by Lukács, although with the difference that sensuous activity is given a more central role and the unit of interconnectedness is thought of not as society in a narrow sense but as the larger ecological whole in which society is embedded. The authors thus advocate a dialectics that, on the one hand, stresses the senses as central to liberation and, on the other, seeks to extend the scope of dialectical mediation to the wider ecology in which humanity is but a part.

Dialectics or Dualism?

We have seen that Foster criticizes Western Marxism for abandoning nature to positivism, and that his strategy for overcoming this is to develop an idea of an ecology-wide dialectics based on natural praxis. This position is close to the position he attributes to Western Marxism, considering the centrality in the latter of the notion of praxis. In my terminology, Foster advocates a practical materialism in which nature is viewed as dialectical primarily because it includes praxis.[2]

However, Foster and his co-authors do not adhere consistently to the conception of a praxis-based dialectics. Sometimes they conceptualize their endeavor as a natural science taking as its object developments capable of unfolding on their own, without human involvement. This can be seen in their frequent reliance on evolutionary biologists or geneticists such as J. B. S. Haldane, Richard Lewontin, Richard Levins, or Stephen Jay Gould to illustrate dialectical approaches in natural science (Clark & York 2005a, 2005b, Foster 2000: 249–54).[3] It can also be seen in recent writings, where Foster and his colleagues strongly assert the existence of nature exterior to humanity, arguing that ecological analysis cannot confine itself to accounting for praxis and that realist and objectivist assumptions must serve as the basis for studying material relations (e.g., Foster et al. 2010: 340ff). That they acknowledge a realm of nature beyond natural praxis and human metabolism with nature is clear also when they approvingly quote the environmental philosopher Richard Evanoff: "natural processes nonetheless can and do continue in the absence of human interaction, suggesting that a measure of autonomy for nature can and should be both preserved and respected" (quoted in Foster & Clark 2016c).

The tendency to acknowledge a dialectics disconnected from praxis is understandable, considering that one of Foster's aims is to show that dialectics is capable of functioning as a natural science, thereby providing the base for a "unity of method." However, his failure to provide any explicit explanation of how to theorize the realm of nature that exists beyond human praxis in a way that breaks with contemplative approaches creates inconsistencies, considering his rejection of contemplative materialism (e.g., Foster et al. 2010: 233f, 247). The only way open if he wants to avoid positing an objective dialectics existing in nature itself would be to emphasize dialectics as a heuristic device for grasping change in nature, but this in itself cannot surmount the contemplative attitude since it doesn't involve praxis. Rather than showing how dialectics ensures methodological unity, Foster reinstates dualism by oscillating between the model of natural praxis applicable to the sensuous way we relate to nature in everyday life and a model of dialectics as a contemplative natural science.

One way to defend Foster is to argue that dualism can be avoided if he can reduce these two models to a common denominator. To a certain extent, he does this when he argues that a common ground uniting the study of nature and society is the use of dialectics as a tool for grasping *mors immortalis*. This move, however, problematically thins out the concept of dialectics, reducing it to a quality that can be studied through a contemplative attitude that goes against the grain of the idea of natural praxis. Making *mors immortalis* a common denominator of dialectics furthermore risks leading to prioritizing such a contemplative attitude also in the realm of society. In view of this, emphatically holding on to the idea of a natural praxis appears to offer better chances of assuring a unity of method, but only as long as nature is thought of in a restricted sense as a nature with which human beings are interacting. A genuine methodological unity, however, would require Foster to include also nature existing outside the orbit of human praxis in the dialectics he is proposing.

To summarize, Foster's search for "unity of method" fails to convince. In his writings he vacillates between two versions of the dialectic, indicating that he ends up reinstating a duality of method contrary to his own intentions. On the one hand, he portrays dialectics as based on a "natural praxis" unfolding in the relation between humanity and nature, which is not in principle different from the praxis-centered dialectics of Western Marxism. To the extent that he foregrounds this version of dialectics, he undermines his own criticism of Western Marxism. On the other hand, he also suggests the possibility of an objective dialectics operative in nature as such. This version of the dialectic brings him close to a contemplative materialism of the kind criticized by Marx. It also suffers from the weakness of having to compete with natural science on its own turf—a drawback that we have seen is avoided in the model suggested by Adorno where natural science is instead accommodated through constellations.

To use the terminology introduced in previous chapters, Foster's texts reproduce the tension between what I have called *practical* and *causal* materialism. His attempt to forge a unity of method takes the form of an attempt to straddle a practical materialism based in the notion of natural praxis and a modified form of the causal materialism of Engels. His inability to surmount the differences between these two positions successfully is shown by the fact that he is forced to recognize the independent existence of a nature beyond the purview of the subject. Ironically, he ends up in a position that is analogous to that of Lukács, the target of much of his criticism. Just as Lukács was unable to show how a dialectics centered on the subject and human praxis could be extended to the domain of nature, Foster is unable to show how the entirety of nature can be encompassed within the realm of natural praxis. Just like Lukács, he is therefore forced to concede that objectivistic methods unrelated to praxis are valid in order to study at least part of nature. Significantly, the Lukács problem reappears in Foster, and his accusation against Western Marxists that they ended up in a dualism thus turns out to boomerang back on himself.

Can We Be Materialists without a Dialectics of Nature?

We can now return to Foster's critique of Western Marxism and scrutinize it closer. I start with the issue of materialism. He argues that Western Marxism ended up in idealism since it jettisoned the idea of a dialectics of nature. Furthermore, its rejection of the dialectics of nature had the "tragic result" that:

> the concept of materialism became increasingly abstract and indeed meaningless, a mere "verbal category", as Raymond Williams noted, reduced to some priority in the last instance ... Ironically, given the opposition of critical, Western Marxism generally ... to the base-superstructure metaphor, the lack of a deeper and more thoroughgoing materialism made the dependence on this metaphor unavoidable – if any sense of materialism was to be maintained.
>
> (Foster 2000: 8)

This characterization of Western Marxism hardly qualifies even as a caricature. Foster's charge disregards that Lukács and other contemporary Western

Marxists didn't reject materialism as such, but only causal materialism. The materialism that Foster attributes to Marx and claims to champion—a materialism emphasizing the role of praxis, including the interplay between humans and non-human nature—was precisely the materialism that most Western Marxists retained. What distinguishes the materialism of Lukács and the early Frankfurt School is not that it is hollowed out or "reduced to some priority in the last instance," as Foster charges. That accusation may be pertinent in relation to the structural Marxism of Althusser, but hardly to Lukács or the Frankfurt School. The same can be said of Foster's criticism that Western Marxists were forced to rely on the base-superstructure metaphor "if any sense of materialism was to be maintained." This remark is particularly off the mark in relation to Adorno.[4] From the latter's perspective, it is precisely reliance on such metaphors that should be criticized as idealism, as an attempt to capture history in the net of reified concepts.

The point at issue here is not which type of materialism represents correct Marxism. The point is rather that other conceptions of materialism exist than Foster's and that he is wrong in claiming that abandoning the dialectics of nature amounts to an abandoning of materialism as such.

Dialectics in Eco-Marxism and Western Marxism

So far, I've argued that Foster is wrong on two scores. Firstly, his attempt to resurrecting a dialectics of nature cannot bring about the unity of method that he aims for. Secondly, rejecting this dialectics of nature does not amount to rejecting materialism per se.

Let me now turn to dialectics in Western Marxism. Just as the rejection of an objectivistic dialectics of nature did not amount to rejecting materialism per se, it did not imply rejecting the attempt to understand nature dialectically. What Western Marxists *did* reject was dialectics in the form of deterministic, objective laws, whether applied to nature or society. Lukács's criticism of Engels's dialectics was not primarily an attempt to limit the reach of dialectics to society but involved a double move. First dialectics was freed from determinism by emphasizing its link to praxis. Only as a consequence of this emphasis on praxis was dialectics, in a second move, viewed as having its home primarily in society rather than nature. However, the withdrawal from nature was not absolute. Theoretically, there is still ample room for theorizing nature in a dialectical fashion in Western Marxism. As we've seen in previous chapters, such an attempt to theorize nature might fruitfully make use of Adorno's ideas of constellations, the primacy of the object, and his discussion of natural history, which shows that drawing a rigid boundary between society and nature is itself undialectical. These ideas make it plain why Foster's accusation that Western Marxism handed over the study of nature to positivism is off the mark, particularly in regard to Adorno.

Neither does it appear to me that anything important is lost by giving up the idea of an objectivistic dialectics of nature. As we recall, Foster suggests that such a dialectics is needed for at least three reasons: it acknowledges the self-consciousness of at least part of nature, it helps us understand its unceasing change (*mors immortalis*), and it does justice to the fact that dialectics always shapes our perception of nature. But these aspects of nature are also taken account of in critical theory. Firstly, the idea of nature as a possible subject is strongly present in Marcuse (1972: 59f) and it also informs the discussion of mimesis in the *Dialectic of Enlightenment* as well as Benjamin's (1988) discussions of nature's language and muteness. Indeed, there is a forceful utopianism built up around the liberation of or reconciliation with nature in critical theory that is much more pronounced than in eco-Marxism (see Chapter 10). Secondly, the idea of "immortal death" is approached by Benjamin (1985) through the lens of the historicity and evanescence of nature. Thirdly, the idea of dialectics as a necessary heuristics for apprehending nature is developed by Adorno, in whose hands it turns into a philosophy guided by the primacy of the object.

But more important than these three points is the central question of how to relate to existing natural science, the so-called Lukács problem. As we recall, Foster claims that the Frankfurt School, by rejecting the objectivistic dialectics associated with Engels, hands over nature to positivism. His own solution is to resurrect a dialectics of nature of Epicurean inspiration. A weakness with this proposed solution is that Foster himself tends toward reinstating a dualism of method, relying partly on a model of natural praxis and partly on more objectivistic forms of dialectical thought as exemplified by evolutionary theory in biology.

We may recall that, as pointed out by Mészáros (1970) and Schmidt (1973: 41f), Marx's ideal of a unitary science presupposes an overcoming of alienation in society. As long as alienation remains, natural science will remain "abstractly material" and "idealistic," as a symptom of estranged needs. Foster by contrast appears to want to realize Marx's vision of a unitary science prematurely by extending dialectics to nature and society here and now. Removing determinism with the help of Epicurean materialism doesn't help. To the extent that he stresses the objectivistic side of dialectics, it reverts to contemplative, traditional theory. To the extent that he instead stresses the subject and praxis, it ceases to be applicable to nature as an objective reality. Since Marx envisioned a unitary science as a utopian undertaking, it is unsurprising that Foster fails to demonstrate its possibility today. But this failure at the same time dissolves much of the rationale for his criticism of Western Marxism, since this criticism strikes back on himself. As I argued in the previous chapter, a more promising way of handling the Lukács problem is offered by the idea of constellations. Rather than aiming for a synthetic stance that would paper over differences for the sake of a premature unity, they bring out the contradictory nature of the object, which is the only way to be true to the fact that society is still unreconciled.

What Can Critical Theory Learn from Eco-Marxism?

The critical theory of nature and eco-Marxism both contribute to understanding the present ecological crisis in relation to modern capitalist society. Both go beyond the rigid dualism between nature and society that has often been a stumbling block in environmental struggles, but they do so in different ways. Is there any possibility of utilizing the respective strong points in the perspectives to make up for shortcomings in the other?

In this chapter, I have discussed to what extent Foster succeeds in demonstrating how a resurrected dialectics of nature can function as the basis for a unitary method spanning both nature and society. I showed, firstly, that Foster's attempt to secure unity of method by reinstating an upgraded version of the dialectics of nature fails. Rather than achieving unity, it tends to operate with two kinds of dialectics, one praxis-based and another contemplative.

Although the former is extended under the name natural praxis to the whole of humankind's ecology, it is not in principle different from the dialectics developed by Lukács and other Western Marxists and his harsh criticism of them therefore seems unfair. Tensions are also introduced in his theory since he supplements this praxis-centered model with seemingly more objectivistic and contemplative conceptions of dialectics, thereby undermining his own attempt to use dialectics as the ground for a unity of method.

Secondly, Foster's charge that Western Marxism abandoned materialism and ended up in idealism ignores the existence of other forms of materialism than his own. From the standpoint of critical materialism, Adorno presents a criticism of idealism that Foster appears to overlook entirely.

Thirdly, Foster's argument that Western Marxism's unwillingness to embrace a dialectics of nature meant that it had to hand over the study of nature to positivism is plainly wrong. The critical theory of nature offers ample resources for thinking nature dialectically (see Chapters 3 and 6). These include the use of constellations and tools for grasping the historicity of nature such as the notions of natural history and second nature. Foster's neglect of these resources is regrettable since Adorno developed his materialism out of concerns not dissimilar to Foster's own, namely, to avoid determinism without jettisoning materialism (Adorno 1974: 256). Adorno did so in a way that one might imagine that eco-Marxists should welcome: far more than Lukács he recognized the autonomy of nature in relation to the subject's attempts to dominate it and the fact that nature exceeds the grasp of the subject.

If we look at the overall theoretical directions of eco-Marxism and critical theory, we can see that both developed in response to problems in practical materialism. In Lukács the category of totality became wedded to a philosophy of history that required the proletariat to play the role of a historical subject. At the same time, the emphasis on the subject meant that he had to struggle with the place of nature in his dialectics. Adorno's response to these problems was to reorient critical theory toward critical materialism. Foster's response, by contrast, was to attempt to conjoin causal and practical materialism. To a considerable extent, the disagreements between eco-Marxism and critical theory therefore appear to express conflicts between underlying varieties of materialism.

Regarding the *politics* of the ecological crisis, the eco-Marxists have strong arguments but the critical theory of nature can defend itself at least in part. Burkett and Foster have argued that Frankfurt School thinkers overlook the possibility that the ecological destruction wrought by capitalism will generate a crisis for capitalism itself and that they neglect the possibility of a unified struggle against economic exploitation and ecological degradation (Burkett 1997, 174, Foster & Clark 2016a). Although I agree that the contradictions generated by the domination of nature deserve more attention, this criticism is only partially true. In the final chapter, I will return to the issue of politics and show how I think this weakness can be remedied.

My criticism of eco-Marxists like Foster and Burkett doesn't mean that I see nothing of value in eco-Marxism. The eco-Marxists have contributed greatly to clarifying how the logic of capital as described by Marx relates to environmental problems, and this contribution should be welcomed from the perspective of critical theory. Foster and Burkett are correct that Marx was more concerned with ecology than Western Marxists and Frankfurt School scholars have usually thought, and it must be conceded that Foster is right that much of the Frankfurt School's criticism of the domination of nature was "carried out without any genuine ecological knowledge" (Foster et al. 2010: 242; see Wilding 2008 for a similar point). This, however, is not fundamentally a theoretical deficiency but rather a problem that can be remedied by more research. Although no Frankfurt School critical theorist has addressed ecological issues as directly as either Foster or Burkett, critical theory certainly possesses the theoretical resources for doing so.

8

World-Ecology and the Persistence of Non-Cartesian Dualism

A significant event in the broader field of eco-Marxism in recent years was the publication in 2015 of Jason Moore's *Capitalism in the Web of Life*. The book presents his highly distinctive "world-ecology" approach, which situates capitalism in a world-historical context based on a meta-theoretical monism in which the interconnection between capital and its outside is conceptualized as a "web of life."

This chapter discusses Moore's work with two aims in mind. The first is to bring out how it helps us move from a critical analysis of the logic of capital to an analysis of how this logic relates to nature. Already here I want to state that this is the part of Moore's theory that I think is strongest and where I think a critical theory of nature has most to learn. In answering this question, Moore makes good use of insights from world-systems theory while anchoring his argument firmly in Marx's value law.

The second aim is to scrutinize Moore's monism and thereby to throw light on the debate in recent years between Moore and John Bellamy Foster around the so-called Cartesian dualism of nature and society.[1] My intention is not to take sides in this rather ferocious debate but to insert critical theory as a third party in it. Conceiving of the debate as a three-way struggle seems reasonable to me since critical theory is already present in it as a point of reference and a target for critique. The primary reason for intervening in this debate is that the stakes in it are high for the critical theory of nature. A monist position seems to undermine the very idea of a domination of nature. I will therefore attempt to map a road forward that steers clear of such monism without ending up in the unresolved tensions of Foster's dialectics. Such a road, I argue, is offered by critical materialism and the form-matter dialectic.

Moore's World-ecology

First we need a look at Moore's world-system inspired approach to ecology. A striking trait is its ambitious attempt to relate capitalism as a logical model governed by the value law to the complex historical reality of the world-system as it has unfolded for the last half-millennium. Marx himself took pioneering steps toward such an analysis with his famous remarks on primitive accumulation in *Capital*, the often brutal and violent appropriation of wealth by non-economical means such as the enclosures and colonialism. From Rosa Luxemburg (2003) onward, a significant number of Marxists have argued that primitive accumulation—or accumulation by dispossession, to use David Harvey's term—is not only a historical precondition for capitalism but an ongoing process and a crucial part of capitalism's regular mode of operation.[2]

A valuable contribution of Moore is that he shows that these processes are not a mere contingent addition to the exploitation of wage-labor but driven by the logic of value-production itself. He presents four important arguments that detail how this happens and what it means for the relation between capitalism and nature.

The first argument introduces his concept of appropriation, which refers to all "extra-economic processes that identify, secure, and channel unpaid work outside the commodity system into the circuit of capital" (Moore 2015: 17). Significantly this involves the appropriation of natural resources. While Marxists have generally emphasized the exploitation of labor as central to capital accumulation, Moore argues that exploitation and appropriation are equally necessary. The two are interwoven, since nature is often appropriated precisely to reduce labor costs, through cheaper food, energy, and raw materials, and to raise productivity. In addition to exploiting wage labor, the system therefore depends on a massive appropriation of unpaid work, including "the unpaid work performed by extra-human nature" (Moore 2015). Moore recognizes that this work, unlike wage labor, does not directly give rise to value, but he emphasizes that it nevertheless contributes by producing the *conditions*—the relations, spaces and energy—needed for value creation (Moore 2015: 29).

Moore's idea of appropriation is based on the notion of primitive accumulation but develops it in two ways. Firstly, he broadens it by stating that it can take place through cultural hegemonies and the development of science and technology in addition to direct coercion. Secondly, rather than focusing on how primitive accumulation serves as a precondition for the creation of the proletariat by depriving people of access to commons and other means of subsistence, Moore focuses on how capitalist production benefits from the material use-values provided by the appropriated resources, which can be used to press production costs and increase relative surplus value. In relation to nature, this means for example that appropriation involves both cheapening costs for resources ("taps") and for waste ("sinks").

To this argument about the system's dependence on appropriation, Moore adds a second argument about how capitalism constantly strives to create or access new "cheap" nature. To understand this argument properly some care is required. For all its unorthodox appearance, Moore's analysis is anchored in the labor theory of value.[3] That means that he believes that since value is created by labor in capitalism, nature has value only to the extent that it is processed by labor. At the same time, he argues that capitalism benefits from "cheap" nature—energy, food and raw materials—and systematically relies on it to mitigate its crises and maintain its profit levels. There is no contradiction between these statements, as we understand when we recall how nature contributes to value. Although nature doesn't create value, it can give individual capitalists a competitive edge against other capitalists and help them increase their share of the aggregate surplus-value produced in society. As Marx points out, increasing the reliance on natural resources leads to a higher organic composition of capital (i.e., a higher proportion of constant to variable capital). While individual capitalists benefit from this in competition since it usually means a higher productivity of labor, the outcome is a downward pressure on the general profit rate. The higher the organic composition, the lower the rate of profit tends to be.

In a significant elaboration on Marx's theory, Moore argues that capitalism always strives to increase its supply of "cheap nature" to prevent the profit rate from falling and that this is one of the primary motors behinds its need for constant expansion. This can sound strange at first, since the increasing reliance on nature is what causes the profit rate to fall in the first place. Moore's argument, however, is that the fall can be prevented if nature can be made cheaper, for example, by discovering new sources of energy or making extraction more efficient. It is thus not just *more* but *cheaper* nature that is needed. In technical terms, cheaper nature means that the organic composition of capital can be kept low while the technical composition of capital (i.e., the proportion of the use-values provided by nature and labor) increases (Moore 2015: 143; see Kincaid 2017). In Moore's terminology cheap nature creates an "ecological surplus" that prevents the profit rate from falling by cheapening production and thereby maintains vitality in the capitalist economy.

> When capitalists can set in motion small amounts of capital and appropriate large volumes of unpaid work/energy, the costs of production fall and the rate of profit rises. In these situations, there is a high world-ecological surplus (or simply, "ecological surplus"). This ecological surplus is the ratio of the system-wide mass of capital to the system-wide appropriation of unpaid work/energy.
>
> (Moore 2015: 95)

For capitalism to maintain its ecological surplus it must constantly expand the available zones of appropriation. Moore argues that this happens through cyclically occurring bursts of world-system expansion that initiate long-term cycles of accumulation. The first wave of European colonization in the sixteenth century that gave the European world-economy access to the resources of the New World is one example. What matters is not just to commodify nature, but to extend "the zone of appropriation *faster* than the zone of commodification" so as to secure a large ecological surplus in the form of large untapped resources of cheap nature for capitalism (Moore 2015: 66).

The emphasis on the continuous production of new cheap natures brings Moore close to a constructivist view on nature. His third argument is that nature is not "just there," waiting to be discovered, but actively produced through science, empire-building, and opening up new markets. Capitalism's reliance on this process means that it systematically reinforces the nature-society dichotomy, which becomes ingrained in modern culture. Referring to Sohn-Rethel, Moore calls it a "real abstraction" that is constitutive of how we organize material reality and as such directly complicit in capital accumulation.

> [B]ecause value was premised on valuing some nature (e.g. wage-labor) and not-valuing most nature ("women, nature, colonies"), it necessitated a powerfully alienating conception of nature as external. At the core of the

capitalist project, therefore, from its sixteenth-century origins, was the scientific and symbolic creation of nature in its modern form.

(Moore 2014a: 17)

As seen in the quote, the "nature" created by capitalism includes large portions of humanity. Referring to how Europeans compared colonized peoples to animals, he asserts that "nature" in the rise of capitalism "*came to include the vast majority of humans within its geographical reach* ... The conquest of the Americas and the paired 'discoveries' of Nature and Humanity/Society were two moments of a singular movement" (Moore 2016c).

This argument throws light on how capital relates not only to class but also to other dimensions of subordination.[4] As Moore puts it: "All manner of racialized and gendered mediations ... have served to normalize the appropriation of humanity's free gifts over the past five centuries" (Moore 2015: 214). By providing unpaid work, or work that is "cheap" in relation to regular wage-labor, humans too are treated like nature, namely, as objects of appropriation that are utilized to press production costs. This production of nature helps us understand why capitalism generates not only class struggles around exploitation but also conflict lines related to appropriation with gender and race as flashpoints. This is a point also made in world-systems theory (e.g., Balibar & Wallerstein 1991: 29–36). Where Moore stands out is in pointing to the continuity between how capitalism utilizes humans and non-human nature.

Moore's fourth argument is that capitalism is running out of cheap nature. This argument shows that his constructionism is qualified. We are now facing "the end of Cheap Nature – and with it, the end of capitalism's free ride" (Moore 2015: 87). In this prediction we hear echoes of Luxemburg's theory that capitalism undermines its own conditions precisely by spreading like wildfire over the globe. As we face not just "peak oil" but also the peak of cheap nature generally ("peak appropriation," as Moore calls it, Moore 2015: 106), it becomes increasingly difficult for capitalism to maintain the ecological surplus and prevent the profit rate from falling. In making this argument, Moore isn't positing an absolute natural limit. Rather, he sees a mixture of several factors behind these difficulties: among them entropy and the depletion of energy and mineral resources, the increasing contradiction between the reproduction times of nature and of capital, protest movements, and the emergence of "negative value"—that is, natures hostile to capital accumulation such as global warming (Moore 2015: 97f, 227f). With the end of cheap nature, capitalism is entering a chronic state of crisis.

Monism from Neil Smith to Jason Moore

While Moore's approach is inspired by and largely consonant with world-systems theory, one of his more provocative arguments cannot be found in that theoretical tradition, namely, his strongly worded rejection of the nature-society dualism in favor of monism. In contrast to such dualism, Moore insists that human and non-human elements are always assembled or "bundled" together in socio-natures. Capitalism itself, Moore writes, isn't exterior to or opposed to nature but "a way of organizing nature." Contrary to the image suggested by Foster's theory of the metabolic rift, capitalism doesn't act *on* nature so much as *through* the "web of life" of which it is part. Rather than focusing on what capitalism *does to* nature, Moore argues that we should focus on what nature *does for* capitalism. In opposition to Foster's "dualism," he therefore advocates a "monist and relational" view of capitalism's metabolism with nature (Moore 2015: 85). The grand dualism of society and nature is replaced by the metaphor of a web of life, which Moore also calls *oikeios* or capitalism-in-nature and nature-in-capitalism.

While explicitly directed against Foster's idea of the rift, Moore's argument by extension also implies a rejection of the Frankfurt School thesis about the domination of nature. Both are stuck in an outmoded dualist framework that pits capitalism against nature. Nature, Moore claims, isn't ravaged or victimized by capitalism, but made to work for it: "capitalism has survived not by destroying nature ... but through projects that compel nature-as-*oikeios* to work harder and harder – for free, or at a very low cost" (Moore 2015: 13).

Moore's unwillingness to think in terms of a domination of nature goes back to Neil Smith, who already in 1980 criticized Alfred Schmidt and the Frankfurt School by arguing that capitalism didn't so much dominate as produce nature—an idea that he developed at length in his influential 1984

Uneven Development and several other texts (Smith 2006, 2007, 2010, Smith & O'Keefe 1980). Smith claims that nature and its processes are constantly reinvented and remade by social production processes. Instead of positing nature as external to society, he emphasizes how for Marx nature was only conceivable as mediated by labor (Smith 2006: xiii). As we have seen, Schmidt too emphasizes this practical or sociohistorical side of Marx's concept of nature. Smith, however, goes a step further and denies that there is any need to think of nature as *anything* but shaped by human production processes. He therefore criticizes Schmidt for unnecessarily clinging to residues of the nature-society dichotomy (Smith 2010: 31–44). The brunt of the attack is directed at Schmidt's equation of nature with use-values. In capitalism, Smith asserts, things are produced for exchange. What we see in capitalism is the production of second nature, the world of exchange, out of the first. But what Schmidt misses is that the first nature is *also* produced:

> Indeed the "second nature" is no longer produced *out* of the first nature, but rather the first is produced by and within the confines of the second. Whether we are talking about the laborious conversion of iron ore into steel and eventually into automobiles or the professional packaging of Yosemite National Park, nature is produced. In a quite concrete sense, this process of production transcends the ideal distinction between a first and a second nature.
>
> (Smith & O'Keefe 1980: 35)

More specifically, Smith argues that nature becomes internal to the logic of exchange-value on the world market. The production of nature is thereby understood as governed by that logic, which Smith theorizes through the notion of uneven development. To Smith, stressing the produced character of nature is politically important since the laws that govern second nature, and thereby also indirectly first nature, are socially created and can therefore be changed and abolished (Smith & O'Keefe 1980: 32). In line with the denial of nature's externality, Smith is also skeptical to the idea of "natural limits" and to apocalyptic arguments in environmentalism which to him appear premised on an untenable opposition between society and nature (Smith 2006, 2010: 243–51). To use Kate Soper's (1995) terminology, Smith comes down as decidedly "nature-skeptical" rather than as "nature-embracing."[5]

Moore shows a kinship to Smith's approach in his stress on the constructed side of the nature-society dualism. Both deny the ontological validity of the nature-society dualism and both downplay or disparage apocalyptic narratives and the risk for catastrophic outcomes. Like Smith, Moore thus rejects dualist viewpoints in part because they easily lend themselves to catastrophist and collapse narratives (Moore 2015: 5). Moore, however, doesn't go as far as Smith in asserting the capacity of capitalism to reshape nature on its own. "Far from asserting the unfettered primacy of capitalism's capacity to remake planetary natures, capitalism as world-ecology opens up a way of understanding capitalism as already co-produced by manifold species" (Moore 2015: 3f). To him, rejecting dualism doesn't mean seeing nature as one-sidedly produced by capitalism. Instead, he sees it as a product of the entire web of life.[6] In his view, agency is possessed not only by living species, but by all bundles of human and extra-human natures, including the climate, weeds, and diseases (Moore 2015: 37). In stressing the agency of the entire web of life, the kinship is not with Smith so much as with non-Marxist approaches such as actor-network theory and new materialism. At the same time, he differs from these currents by retaining and attempting to develop Marx's value law and his stress on dialectics.

The Foster-Moore Debate

In *Capitalism in the Web of Life* Moore criticizes Foster's idea of the metabolic rift for reinforcing a "Cartesian dualism," according to which humanity is seen as existing apart from nature (Moore 2015: 75–87). This dualistic conception, he claims, prevents Foster from realizing the promise of a unitary method that was inherent in his early groundbreaking formulations of the idea of metabolism. The problem is that the idea is forced into a dualist script, a "kind of ping-pong between 'natural forces' and 'human agency'" that rested on human exceptionalism (Moore 2016b).

In reply, Foster has repeatedly charged Moore as well as proponents of the production-of-nature approach like Smith with furthering a monist ontology that denies any contradiction between capital and nature (e.g., Foster 2016a: 393). The remedy, he argues, is returning to Marx and reasserting him as a model for "reintegrating the critique of capital with critical natural science"

(Foster & Clark 2016c). Political invectives have not been missing in this debate. Foster is thus concerned with what he sees as the baleful influence of non-Marxist approaches such as actor-network theory on Moore's theorizing.[7] In Foster's view, Moore ends up in a position that is no longer Marxist but idealist and incapable of properly diagnosing the ecological crisis:

> The end result is pure idealism ... To say, as Moore does, that considering natural processes on one hand and capitalist valorization processes on the other is a dangerous dualism ... eliminates the very possibility of an ecological critique of capitalism ... How are we to judge an analysis that excludes the ecological movement and its perspectives, abandons Marx's value analysis, has nothing at all to say about class struggle, and leaves humanity's fate to the evolution of capitalism as a singular, bundled actor? ... Moore, I am sorry to say, has moved to the other side, and now stands opposed to the ecosocialist movement and socialism.
>
> (Foster 2016b)

In reply, Moore denies that a monist position must lead to a loss of differentiation:

> Nothing could be farther from the truth! Seeing human organizations as part of nature leads us to explore manifold socio-ecological connections that make us specifically human – just not "exceptional." ... [I]t will be hard to develop a politics of emancipation for all life without a philosophical commitment to precisely that: emancipating *all* life. And an authentically multi-species politics of emancipation will require – and will need to nurture – ways of thinking that connect first, and separate later.
>
> (Moore 2016a)

Both Foster and Moore claim to champion dialectics. According to Foster (2013), dialectics necessarily involves the temporary isolation of moments within a totality for the purpose of analysis. The idea of the metabolic rift is therefore not dualistic but dialectical—this in contrast to Moore's talk of "bundling" which removes dialectical tension between society and nature with the result that Moore becomes insensitive to ecological crises (e.g., Foster 2013, 2016a: 414). For his part, Moore similarly asserts that dialectics is about the proper deployment of abstractions. What is needed is to search out adequate abstractions based on a sensitivity for historical context rather than to conceive of history as a ping-pong game between society and nature.

Nature and society, he claims, are "undialectical" and "violent" abstractions that take the established Cartesian nature-society duality for granted. That duality is itself a product of capitalism, a result of the latter's need for cheap nature (Moore 2015: 76, 2016c, 2017). To Foster, then, a monist account blunts the instruments needed to criticize capitalism, while to Moore Foster's analysis is caught in abstractions produced by capitalism itself.

Although my aim is not to adjudicate in this debate, it can safely be said that several of Foster's claim are wrong or exaggerated. It is hardly correct that Moore abandons Marx's value analysis. Neither is it fair to dismiss his approach as undialectical or lacking in power to criticize capitalism. Rather than committing to an undifferentiated monism, he clearly recognizes that the nature-society duality is a historical fact that cannot be wished away. On the other hand, Foster's approach cannot simply be labeled dualist. Indeed, we have already seen that he is inspired by a form of monism—Epicurean materialism—and that his chief aim is to use dialectics to surmount the divide between society and nature.

Despite Moore's self-labeling as a monist, I will argue below that what's most valuable in his analysis of capitalism is compatible with a *methodologically* motivated dualism. Such a dualism does not require that the categories of nature and society are granted ontological validity, but it still allows us to make arguments about "rifts" as well as about a domination of nature. The dualism that I propose also lets us to see how the categories of nature and society are undermined and destabilized by their contact with matter, the non-identical object that they claim to cover. The latter point is important since it enables us to escape what Moore describes as the ping-pong match between society and nature.

To make this argument, I will proceed in two steps. I will first return to Smith's production-of-nature approach and assess it from the point of view of critical theory. I will then develop my argument into an argument for why critical theory needs a dualism of form and matter. Such a dualism is vastly different from the Cartesianism criticized by Moore. In the final part of my argument I return to Moore, arguing that such a dualism is superior to monism.

The First Step: A Criticism of Smith

Smith's argument that the idea of a domination of nature should be replaced with the notion of a production of nature amounts to a far-reaching constructivism, but with a qualification that deserves to be highlighted: "To say that nature is produced does not imply that every atom of some tree, mountain or desert is humanly created ... It does mean the human activity is responsible to a greater or lesser extent for the *form* of matter" (Smith & O'Keefe 1980: 36; my emphasis). Smith here brings in the distinction between form and matter in a way that reconceptualizes the nature-society distinction and indicates the limits of his constructivism. While form is social, matter is not necessarily so. Matter here becomes a stand-in for what many would call nature, although Smith denies it this term.

The form-matter distinction brings out the contrast between Smith and the critical theory of nature. As we have seen, Adorno too focuses on the relation between form and matter, or concept and object, but, unlike Smith, he constantly highlights the tension or non-identity between the two. What he calls the primacy of the object implies that matter is never entirely captured by form. By contrast, Smith's attempt to capture the relation between society and nature one-sidedly from the perspective of a social production of nature presupposes that form is unimpeded by matter. In his account, matter vanishes into form, just as use-value vanishes into exchange-value. Against such a position, Schmidt and Foster are right to insist that processes of metabolism cannot be reduced to the form imposed by the logic of capital. While capitalism certainly *aspires* to such a reduction, mistaking that aspiration for reality overlooks the contradictions that this aspiration generates.

We can now reassess Smith's criticism of Schmidt. In defense of Schmidt, a certain kind of dualism between society and nature is perfectly defensible. That dualism is not based on the opposition between two forms, society and nature, but on the opposition between form and matter. As Smith points out, the forms through which we apprehend the world are all socially mediated. This is not a controversial statement; it echoes Lukács's statement that nature is a social category and dovetails with Adorno's idea of natural history, which is a tool for tracing this social mediation. What Smith misses, however, is that form doesn't rule supreme. To be sure, he admits that not all consequences of the social production of nature are intended (Smith 2007: 24f). But by presuming that these consequences can be adequately captured from the perspective of form, he nevertheless denies them any important theoretical role.

We have seen that both Smith and Schmidt emphasize the practical side of Marx's concept of labor—the fact that we know nature only by entering into relation with it. But to Schmidt, this doesn't prevent us from recognizing an irreducible otherness or objectivity in nature. We know it through our praxis, our metabolism with it, on which we depend as living beings, but it is never identical to our representations of it. Stressing the merely social character of nature reduces away what is objective in it. What Schmidt describes here as objectivity is also what prevents him from collapsing first and second nature into each other, as Smith does. By missing out on this objectivity, or non-identity, Smith also misses an important site where critical impulses arise. By overemphasizing the process whereby first nature is incorporated into second nature, he loses sight of the contradiction between use-value and exchange-value, and between metabolism and reification. The consequences of this loss can be seen in his truncated division of the world into two natures—first and second—with no mention of history. When Benjamin or Adorno discussed the preponderance of second nature in capitalist society, they always dialectically interrelated this second nature to history and the possibility of reawakening agency. History is precisely the point where people are empowered against the natural semblance to attempt to bring about change, the qualitatively new.

In criticizing the idea that nature is wholly produced, we shouldn't commit the opposite mistake of suggesting that it is wholly autonomous. Smith's constructivist approach to nature is valid for the realm of social forms. In that realm it would certainly be ideological to uphold any absolute (or "Cartesian") nature-society duality, as Benjamin's and Adorno's ideas of natural history show. Due to Smith's neglect of the primacy of the object, however, he goes wrong when he reduces the dynamics behind the production of nature entirely to the logic of exchange-value. Such a one-sided reduction amounts to a neglect of the contradiction or non-identity between form and matter.

Step 2: Returning to Moore

The dualism of form and matter should not be confused with a Cartesian dualism that posits nature as soulless and inert. Environmentalists are right to reject that duality. But we need tools for grasping the effects of capitalism on both humanity and non-human nature. That requires us to keep our eyes on the relation between capitalist forms and the matter on which they are imposed. The fundamentals of such an approach have, as I have argued, been developed in rough outlines by Adorno. It consists on the one hand in the primacy of the object, which implies attention to the non-identity between concept and object (or form and matter). On the other hand, it consists in tracing the forms through which nature appears to us, along the lines suggested in the idea of natural history. When we trace these forms, we make use of the established nature-society duality but turn it into an object of immanent criticism rather than presupposing it as an ontological fact. Borrowing a term from Kohei Saito (2017b: 289), this approach is a variant of methodological dualism that starts out with a dualist framework for methodological, rather than ontological, reasons.

Foster is right in rejecting monism to the extent that a monist position challenges the idea of a domination of nature or of a metabolic rift. But unlike him, I want to salvage the valuable parts of Moore's theory, above all his world-systems theoretical insights into processes of appropriation that I do not see as depending on monist premises. I will therefore suggest that his approach is not as monist as he himself claims. His professed monism cannot hide that in practice much of his analysis proceeds as dualistically as those of his eco-Marxist critics, starting out from the logic of capital and then studying how this logic relates to the rest of the web of life. In fact, his adherence to a Marxian explanatory framework requires him to give the capitalist economy a certain autonomy from the rest of the web of life, since its logic is treated as a fixture that subsists over time in isolation of its interaction with the rest of nature. His inability to surmount dualism is reflected in the fact that he operates with two ideas of nature: nature as produced by capitalism and nature as a "web of life." That he pulls back from claiming that the latter is entirely produced by capitalism means that a nature-society divide remains in his writings.[8] The resulting picture is most convincing where it goes against the grain of its own monism.

A form-matter dialectics provides, I suggest, a platform for integrating the best insights of Foster and Moore, insights that would include both Foster's discussions of the metabolic rift and Moore's analysis of appropriation. The integration is facilitated by that fact that both believe that the nature-society duality is a historical creation that shouldn't be ontologized. Foster tries to bridge it with a dialectics of nature, while Moore adopts a monist vantage point from which to explain its historical emergence. The form-matter dialectics steers clear of the weaknesses of Foster's brand of dialectics as well as Moore's professed monism. What matters is therefore not to choose between their respective approaches, but rather to assert what a critical theory can contribute in comparison with them. Let me highlight these contributions in regard to three points.

The first point concerns Smith's and Moore's polemic against the notion of a domination of nature (or of capitalism acting "on" nature rather than "through" it, as Moore puts it). If nature cannot be viewed as wholly constructed, then this polemic cannot be sustained. Their arguments about the production of (cheap) nature and the creation of the nature-society divide as a real abstraction don't contradict the thesis of dominating nature. What they describe with these terms is the production of *forms* that capitalism imposes on nature, and such imposition is a form of domination.

The second point concerns how to criticize these forms. A critical materialism asserts that it is crucial to focus on the relation between form and matter. That means that we should avoid the blackmail of choosing between Cartesian dualism and its monist negation, since both of these options operate with social forms ("nature," "society," or some mixture of these). Critical theory, by contrast, seeks to *subvert forms* through the experience of non-identity. As Horkheimer asserts, not only dualism but also "every kind of philosophical monism, serves to intrench the idea of man's domination of nature" (Horkheimer 2013: 169). When criticizing the Cartesian dualism, we should not elevate monism in its stead.

The final point concerns the political implications of the form-matter dialectics. When Smith and Moore stress how nature is produced, they do so in part for political reasons: by showing that nature is historical they demonstrate that it can be changed. The dualism of form and matter that I propose in no way denies the changeability of nature. Such dualism doesn't reify nature by endowing it with an ahistorical essence, but instead seeks to dereify it. Nature is changeable but attempts to change nature must go hand in hand with recognizing that it is never identical to the forms we impose on it.

Conclusion

By way of ending I want to stress both the strengths and weaknesses in Moore's approach. His theoretization of capitalism's dependence on its outside in terms of appropriation is valuable. Rather than representing a deviation from value theory, as Foster claims, it explores and contributes to a clarification of how capitalism mobilizes use-values and how these interrelate with the production of value. By engaging in this exploration, Moore helps us see how capitalism—in addition to the official economy of capitalization, commercialization, and wage-labor—relies on a zone of appropriation of use-values created outside the circuit of labor power, for instance, the unpaid work of households, the non-patented knowledge of indigenous communities, and, of course, nature. Investigating these layers of capitalism demands an openness for combining a Marxist value-form analysis with empirical and historical research. Moore helps us do that, but, as I have argued, his contributions can be adopted without the monist framework in which he presents them.

Regarding the issue of monism and dualism, I concur with Foster that we need a dialectical approach capable of highlighting the contradiction between capitalism and the rest of nature. My solution, however, is different from his. Rather than building on the idea of a dialectics of nature I propose a dialectics centered on the relation between form and matter. This is unlike Cartesian dualism, which Moore and many environmentalist thinkers are right to reject. Cartesian dualism ontologizes the separation between nature and society, leading to the reification of both. As Moore points out, capitalism has turned nature and society into real abstractions, meaning that it systematically reproduces them as the *forms* through which we apprehend matter. But resorting to monism to overcome these forms is premature. Since the separation between nature and society is embedded in social relations, abolishing it is possible only by changing these social relations, by praxis. It can be abolished neither by theoretical *fiat*, as in monism, nor by bridging the gap prematurely by a dialectics of nature, as in Foster. As championed by Smith and Moore, monism is problematical since it prevents us from talking of a domination of nature. Critical materialism is preferable since it allows us to capture the dialectics of form and matter, showing how the imposition of form is an essential part of domination while at the same time criticizing the forms immanently.

9

New Materialism and Dark Ecology

The recent decade has seen a general surge of interest in materiality, matter, and objects, sometimes referred to by speaking of a material or ontological "turn" in social theory.[1] Although this turn encompasses a wide variety of approaches, they share an interest in theorizing the independent agency of matter and in questioning the dichotomy of society and nature. What are the strengths and weaknesses of these approaches and what can be learned from them? In this chapter I want to reflect on this question by focusing on the philosopher Jane Bennett and the literary scholar Timothy Morton—associated with new materialism and object-oriented ontology respectively—who have both drawn in part on critical theory while working out novel ways of addressing environmental issues without relying on a nature-society dichotomy.

Looking into these approaches is especially relevant when considering the theoretical implications of the increasing entwinement of society and nature in today's world. Part of the background of the mounting criticism of the nature-society dichotomy is the recognition that society and nature have become so enmeshed that it hardly makes sense any longer to distinguish the two. I will contextualize my discussion by focusing on this entwinement, which has been referred to as a justification for the criticizing the Frankfurt School idea of a domination of nature. Such a contextualization will bring out not only how the new approaches have responded to that increasing entwinement, but also in what sense the entwinement challenges the idea of the domination of nature.

My argument will be that, while these approaches open our eyes to new or insufficiently explored ways in which human and non-human nature interpenetrate, they build on assumptions that obscure the harm capitalism does to nature. Often being at least in part inspired by actor-network theory, they have been animated by a suspicion or rejection of macro-theoretical theorizing in general and Hegelian-Marxist notions of totality in particular.[2] The stress has usually been on contingency, nominalism, and the micro-level in opposition to the determinism, conceptualism, and predilection for macro-level analyses associated with Hegel and Marx. As I will show, their reluctance to engage in macro-level theorizing hampers their ability to identify the role of capitalism in environmental destruction.

The approaches associated with the material and ontological "turns" are sometimes described as monist and as characterized by flat ontologies. They are indeed monist since they dismantle the ontological separation between society and nature, and their ontology is flat in the sense of not privileging any specific ontological realm (e.g., society or the human mind) with explanatory properties such as agency. At the same time, care is needed when applying these labels. Monism and flatness deny that we can reduce appearances to a separate, deeper level of reality, but do not necessarily imply an uncritical acceptance of surface appearances. Morton, for instance, refers to sensations of non-identity to disrupt such appearances without having to revive the nature-society dichotomy. To the extent that matter or objects are seen as brimming with life, they are also seen as mutually opaque, spectral, and unpredictable (Harman 2018: 11f, Morton 2017: 12).

My argument is not that these approaches are uncritical, but that the critique is one-sided. They rely on immersion in the vital or spectral world of matter or objects to criticize or subvert our concepts and dichotomies but tend to adopt an uncritical and frequently celebratory attitude toward this world itself. To adopt a critical perspective on this world, and the role played by capitalism in producing it, immersion must be supplemented by distance and a proper deployment of macro-level theoretical concepts. Since environmentally destructive mechanisms are ingrained in the world of matter or objects itself because of the imposition on it of capitalist forms, it is highly problematic to celebrate the vitality of that world. By the same token, criticizing such destruction cannot be limited to a critique of macro-level concepts but must take aim at those forms.

That a monist position can undermine critique is an argument that we recognize from the discussion of Jason Moore in the previous chapter. Moore, however, has a theory of capital. Thanks to his ability to contrast the logic of capital to the rest of the web of life, he can conduct his analyses in a way that is methodologically dualist in fact if not in letter. This road is barred to Bennett and Morton due to their one-sided emphasis on immersion. In contrast to them, I will argue for the advantages of holding on to a *critical* materialism that combines a Hegelian or Marxist emphasis on totality with an attempt to criticize that totality immanently.

Vital Materialism

The new materialism of Bennett's influential book *Vibrant Matter: A Political Ecology of Things* (2010) differs from the varieties of "old" materialism that have usually dominated interpretations of Marx. Dialectics and totality, for instance, do not play any role in it. Instead, it stands out by its attempt to dissolve dichotomies such as mind and matter, or humanity and nature. Humans are seen as material beings while matter is endowed with agency and vitality.

There is much to admire in Bennett's work, perhaps above all the attempt to present the world as it would appear to us without the dominant categories dividing us into humans and non-humans and the reification that makes us see both as nothing more than things. As she writes, she tries to reinvoke "childhood experiences of a world populated by animate things rather than passive objects." Poetically, she recognizes that the vitality of matter is hard to discern, since it is "as much wind as thing" (Bennett 2010: 119). Much in this attitude is reminiscent of what Horkheimer and Adorno call *mimesis*, the relaxation of self-control that allows us to react spontaneously to our surroundings without concern for self-preservation. Their stress on the need for human beings to break the spell of instrumental reason by recognizing themselves as nature is a further point of commonality with Bennett. Important is also her acknowledgment of the normative implications of recognizing the vitality of matter. She advocates it because "the image of dead or thoroughly instrumentalized matter feeds human hubris and our earth-destroying fantasies of conquest and consumption" (Bennett 2010: ix). Recognizing the agency of things is a way to "promote greener forms of human culture and more attentive encounters between people-materialities and thing-materialities" (Bennett

2010: x). Conversely, human beings need to reflect on the limited nature of their own agency, which implies care and modesty in relation to things.

Throughout the book, Bennett infuses matter with vitalist connotations such as life, growth, contingency, unpredictability, and becoming while rejecting the idea of matter "as passive stuff, as raw, brute or inert" (Bennett 2010: vii). Like in actor-network theory, matter is seen as possessing agency, or "thing-power," defined as "the capacity of things—edibles, commodities, storms, metals—not only to impede or block the will and designs of humans but also to act as quasi agents or forces with trajectories, propensities, or tendencies of their own" (Bennett 2010: viii). Agency is not necessarily tied to intentionality. It exists not only in individual objects—such as trash or crystals—but above all in assemblages, such as an electric power grid. Agency, she claims, is distributed across a heterogeneous field rather than being located in a single body. This is true also of human agency, which is less an individual capacity than an assemblage.

> The sentences of this book also emerged from the confederate agency of many striving macro- and microactants: from "my" memories, intentions, contentions, intestinal bacteria, eyeglasses, and blood sugar, as well as from the plastic computer keyboard, the bird song from the open window, or the air or particulates in the room.
>
> (Bennett 2010: 23)

Not only non-human things but also the materiality of human existence—the minerality of our bones, the metal of our blood, the electricity of our neurons—make up the assemblages that are the source of agency. "Each human is a heterogeneous compound of wonderfully vibrant, dangerously vibrant, matter" (Bennett 2010: 12f).

Her affinity to critical theory is underlined by her repeated stress on the similarity between thing-power and what Adorno calls non-identity and the primacy of the object (e.g., Bennett 2010: 13–16). Following Adorno, she points out that thing-power isn't outside human experience, "for Adorno describes nonidentity as a presence that acts upon us: we knowers are haunted" (Bennett 2010: 14). She refers to negative dialectics as the "pedagogy inside Adorno's materialism" that helps us train ourselves to detect and accept nonidentity.

This pedagogy involves the self-criticism of conceptualization, a utopian imagination, and a playful element (Bennett 2010: 14f).

However, there are also distinct differences between her approach and critical theory. She herself points out one difference by rejecting demystifying critique (Bennett 2010: xiv). When she rejects such critique as solely aimed at uncovering human agency, she has in mind a critique premised on an opposition between reified forms and human agency. However, as I have argued, reification should not be opposed to the human but to the historical. It is the ahistorical appearance that things possess an innate essence independently of historical context that constitutes reification. That is why even labeling a person "human" as opposed to things or animals can be an act of reification. With this understanding of reification, nothing says that a demystifying critique must stop at uncovering human agency.

My objection against new materialism is therefore not that it endows nature with vitality or agency. A critical theory of nature denounces the oppression of all life, including that of non-humans. But unlike new materialism, it insists that we need macro-level concepts like capitalism to understand that oppression. Here a more serious gulf between Bennett and critical theory appears. Under the rule of identity perpetuated in capitalism, it is not enough to simply promote a mimetic attitude, which would uncritically accept the products of capitalism along with everything else. A celebratory account of vitality in general would have to be extended to the thing-power of capitalism, or, as Christopher Buck (2013: 129f) points out, to the enchanting aspects of commodity fetishism. An additional risk is that Bennett's vital materialism, by assuming everything to be vibrant, misses how humans as well as non-humans can be deprived of vibrancy by being reified. While Bennett is good at tracing the emergence of agency in assemblages and illuminating the vitality of objects that is easily overlooked when the latter are simply subsumed under concepts, we also need to direct attention to what is not vital: second nature in the sense of petrified and reifying structures. Like actor-network theory, new materialism revels in following actors and networks and incorporating non-human nature in enlightening analyses, but without a concept for capitalism as a system of real abstractions it is hard to adopt a critical perspective on the agency of things.

Dark ecology and Hyperobjects

Morton more explicitly than Bennett seeks to develop tools for thinking macro-level objects. Before discussing whether this offers a more promising ground for theorizing second nature, we must point out the considerable ground that he shares with Bennett: the desire to abolish the distance to the object that comes with the presumption of being a detached observer, the injunction to immerse ourselves in the vibrant flow of the object-world, and the attack on the nature-society duality. These points are connected since it is because the duality prevents immersion that he wants to eliminate it. In his 2007 work *Ecology without Nature*, he states that he wants an "ecology without the concept of the natural" (Morton 2007: 24), since the idea of nature is getting in the way of properly ecological forms of culture. At the heart of the problem with such a concept of nature, he states, is distance—the separation that the concept induces between the subject and its environment, a distance reproduced not only in science and industrial technology but also in romantic aesthetics (Morton 2007: 24–8).

Unlike new materialism, Morton does not downplay criticism in favor of affirming the vitality of flows and assemblages, but insists that his purpose is to forge tools for ecological criticism, or "ecocritique" (Morton 2007: 6, 8). While rejecting the nature-society dichotomy, he is therefore equally critical of monist approaches that lead to uncritical accounts: "Monism is not a good solution to dualism," he writes. "Idealism and materialism can both generate flat worlds in which there is no otherness" (Morton 2007: 151). For a solution, Morton turns to Adorno's notion of non-identity, which he describes in terms of slime and the abject, as "shit" that disrupts the semblance of flatness and confronts us with otherness. The guiding slogan, he explains, is not to be "afraid of nonidentity" (Morton 2007: 13). The beautiful soul must be confronted with its shit and made to recognize itself in it. "Ecological art is duty bound to hold the slimy in view. This involves invoking … the pulsing, shifting qualities of ambient poetics, rather than trying to make pretty or sublime pictures of nature" (Morton 2007: 159f).

Morton's embrace of the abject leads to what he calls "dark ecology," described as an affirmation of queerness and as an encounter with radical non-identity. As an example of what we need, he compares to Benjamin's exploration of the melancholy Baroque tragic drama: "Isn't this lingering with something painful, disgusting, grief-striking, exactly what we need right now, ecologically speaking?" (Morton 2007: 197). He also sees it as embodied "in a 'goth' assertion of the contingent and necessarily queer idea that we want to stay with a dying world" (Morton 2007: 184f). He describes it as preserving "the dark, depressive quality of life in the shadow of ecological catastrophe" and exemplifies it with the "undigested grief" that pervades the film *Blade Runner* (Morton 2007: 187). Refusing the romantic ideal of a reconciliation with nature, dark ecology brings about the depressive realization that we are inextricably enmeshed in the decaying, polluted world of human-natural hybrids. "Ecological politics has a *noir* form. We start by thinking that we can 'save' something called 'the world' 'over there', but end up realizing that we ourselves are implicated" (Morton 2007: 187).[3]

In contrast to the celebratory tendency in Bennett, Morton's dark ecology is dyed in a post-apocalyptic sensibility. The problem of nature "has more in common with the undead than with life," he writes. "Environmentalism cannot mourn the loss of the environment, for that would be to accept its loss ... The task is not to bury the dead but to join them" (Morton 2007: 201). His harshest criticism is reserved for Romantic ecologism with its impossible striving for natural purity and separation: "I'd rather be a zombie than a tree hugger" (Morton 2007: 188).

If Bennett tends to miss what is *not* vital, namely, the world of reification, Morton makes a concerted effort to show that the seemingly dead, reified world too is alive—although with queer, slimy, or abject lifeforms that seem to signal the end of humanity, such as the zombie or replicant. We might wonder why this is a useful model for ecocritique. To be sure, it disrupts our idealist constructions of nature and the stance of affirming the abject is perhaps liberatingly honest. But something is problematical. In caring for our shit, shouldn't we still make a distinction between taking responsibility for limiting it and helping it proliferate even further? What in Morton's deep ecology prevents us from giving our blessing to pollution? Affirming shit sounds like affirming toxic waste. In addition to facing our shit, shouldn't we also face the system that produces it?

It is hard to find straightforward answers to these questions in Morton's work and he himself acknowledges that his critique is entangled in the Romanticism he castigates (Morton 2007: 26, 142). The ambiguity seems to me to indicate a weakness in his strategy of immersion. Facing non-identity to disrupt flatness is fine, but why must we discard the concept of nature to do so? Adorno held on to it, not because he took it literally but because it was useful for at least making a gesture toward what was victimized by the system he was criticizing. Referring to Adorno, Morton asserts that "[t]he final word of the history of nature is that nature is *history*" (Morton 2007: 21). But this is incorrect. Asserting the historicity of nature is only one side of the idea of natural history, but Adorno also asserts the opposite: the constant historical pressure for the category of nature to be reproduced. Rather than abolishing the category of nature, he applies it dialectically, in its shifting relationship to history, thereby making it available for him as a category to strategically defend, when needed, in opposition to history.

In his following works, Morton addresses the need to create concepts for macro-entities by presenting the ideas of hyperobjects and subscendence. Hyperobjects are defined as "things that are massively distributed in time and space relative to humans" (2013: 1). Exemplifying them with black holes, global warming, mass extinctions, oil fields, and the Everglades, he stresses that they are not just mental, but real entities whose reality is withdrawn from humans. Due to their scale, they disrupt our sense that we are in charge of endowing meaning on the world. There is a posthumanist sensitivity in Morton's recognition that they humiliate us. "The Titanic of modernity hits the iceberg of hyperobjects … This book is part of the apparatus of Titanic, but one that has decided to dash itself against the hyperobject" (2013: 19). The idea of hyperobjects is deployed as a stand-in for the non-identical, as part of a method whereby concepts are shipwrecked on the non-identical object.

If concepts cannot grasp the hyperobject, how can we form an idea of it? Morton's answer is subscendence. "We need holism, but a special, weak holism," he states (2017: 103)—a holism that no longer believes that the whole is greater than the parts. Against that he posits wholes that are brittle, fragile entities that are smaller and weaker than their parts: "Tim Morton is so many more things than just 'human'. A street full of people is much more than just a part of some greater whole called 'city' " (2017: 101f). Such wholes are less than their parts and characterized by subscendence. This weak form of holism is needed to think hyperobjects—things that are thinkable and computable, yet impossible to see such as global warming, evolution, and extinction (2017: 105).

So here is a way to think macro-objects. But is it also a way to forge an effective conceptual apparatus for criticizing ecological destruction? In *Dark Ecology* (2016), Morton turns his critical attention to what he calls agrilogistics, a "gigantic machinery" which was born in the Fertile Crescent twelve million years ago and which went on to generate today's global warming and mass-extinctions. In line with his idea of subscendence, this hyperobject is not portrayed as monolithic but as a series of "nested catastrophes" described as loops within the loop of agrilogistics (Morton 2016: 42–59, 69ff). Rather than succeeding each other in time, they form "concentric temporalities" of ongoing catastrophes, allowing us to focus in on those that we can do something about (ibid. 77). It is still not clear, however, what he wants to do about present-day catastrophes. Nor is it clear how agrilogistics relates to present-day capitalism. Let us look at his 2017 book *Humankind*, a substantial part of which is taken up by a discussion of Marx and a reinterpretation of commodity fetishism. As Marx points out in a famous passage, fetishism makes things assume a life of their own.

> The table is made of wood, an ordinary sensuous thing. But as soon as it emerges as a commodity, it changes into a thing which transcends sensuousness. It not only stands with its feet on the ground, but, in relation to all other commodities, it stands on its head, and evolves out of its wooden brain grotesque ideas, far more wonderful than if it were to begin dancing of its own free will.
>
> (Marx 1990: 163f)

In keeping with the premise that all beings have agency, Morton argues that Marx was wrong to think of dancing tables as strange. To hold that objects have agency, he asserts, is not an instance of commodity fetishism, but correct regardless of capitalism (Morton 2017: 59f). Things normally dance, or as he also writes, they are spectral—characterized by unpredictable oscillation, wiggling, and swaying to and fro, without any external motive force (Morton 2017: 48ff, 54ff, 178).

In this argument, we recognize an affinity not just to Bennett's vitalism but also to her rejection of demystifying critique. Just as she argues that such critique obfuscates thing-power, Morton argues that the critique of commodity fetishism obfuscates the fact that all things dance. Rather than seeing dancing as the fruit of fetishism, he presents it as an innate quality in all things. In a good formulation he writes: "Commodity fetishism isn't about just the alienation of humans, but the alienation of any entity whatsoever from its sensuous qualities" (Morton 2017: 60). I can assent to this. It agrees with my argument that reification isn't to treat humans as dead things, but to treat anything, including nature, as a dead thing.

However, what's missing in Morton is precisely an analysis of reification. It's true that nature is full of agency. But reification means that it's treated as a dead thing, which is *then* made to dance according to the tune of capitalism. *That* is the dance that Marx criticizes. The reality of commodity fetishism shouldn't be brushed aside by the assertion that everything dances even without it. Such an assertion blinds us to what is problematic with commodity fetishism. It is hardly surprising that Morton's ecocritique seems to contain no ambition to criticize the system from which this fetishism emerges. Instead he suggests that we should let things rock and dance: "We can only fix our problem by allowing things to move all by themselves … Which in turn, means that we need to let tables dance … Actually, need to let tables rock" (Morton 2017: 177). This upbeat exhortation misses what's typical of the dancing produced by capitalism. Fetishism doesn't just mean that things dance; it means that the social context of their production is obscured by the exchange-relation on the market. To fix *that* problem, we need to find a way to let things dance to other tunes than the one set by capitalism. While letting capitalism slip out of the analysis, Morton urges us to immerse ourselves in the spectral world to rediscover solidarity with nonhumans. Marxism, by contrast, points to what structurally obstructs solidarity.

Two Usages of Concepts

Both Bennett's new materialism and Morton's object-oriented ontology usefully direct attention to the agency of matter or objects, drawing on their non-identity with established concepts to criticize the divisions imposed by the latter. My trouble is not with the celebration of vibrant flows or the *noir* revelation in slime and non-identity, but with the neglect of capitalism. This neglect is not surprising considering the impact on these approaches of poststructuralist thought and actor-network theory—currents of thought that are infused by a strong anti-Hegelianism and suspicion of the notion of totality.

To clarify why I believe that capitalism can be usefully described as a totality, I want to make a distinction between two ways of using concepts. The first judges concepts based on how well they are supported by empirical facts. The idea that concepts should correspond to facts is characteristic of what I have called causal materialism. To a considerable extent, this idea underlies the dissatisfaction with "totality" in the approaches discussed in this chapter, but what is overlooked in this idea is that concepts need to simplify in order to carry meaning at all—that is, to help us grasp what is essential about an object. This is why attention to the material side of objects can always reveal them to be non-identical to their concepts. The attempt to pursue correspondence has a paradoxical quality since it tends to undermine meaning and ultimately, as in nominalism, empties out meaning from concepts altogether.

There is, however, another way of viewing concepts that makes the notion of totality more defensible. Totality can be viewed as a methodological rather than as ontological concept—as a deliberate fiction constructed by thought in order to create meaning. Marx's model of capitalism is such a totality, a reconstruction of the constitutive moments of the concept of capital. In that sense, it functions like a Weberian ideal type, a heuristic tool that is helpful even while it is recognized not to correspond to the complexity of reality. Without setting up totalities of this kind, there can be no emphatic grasp of the phenomena we observe, since it is totalities that create meaning. Such totalities don't have to involve any presumption that we have arrived at absolute truth, that there is no outside to it or that we put ourselves above facts. It's only if we insist that concept and reality do in fact correspond that totality becomes totalitarian, a synonym for conceptual closure. At the same time, recognizing the fictional quality of totalities doesn't mean that "anything goes," since it is the felt necessity of making sense of a matter, such as a certain experience or problem, that impels us to totalize.

Neither Bennett nor Morton appears to recognize the ideal-typical use of concepts. To some extent they make up for this by introducing ideas—

such as assemblage, weak holism, and subscendence—that indicate ways of approaching complex macro-entities without subsuming them under concepts. But unlike ideal types, these ideas stress fidelity to empirical flows and interconnections rather than conceptual clarity and integrity. Assemblages, for instance, are contingent orderings in which individual elements retain their freedom to operate as independent actants. Like Morton's weak wholes, they enable us to gesture toward macro-level wholes without totalizing (see e.g. DeLanda 2016: 9–12). Even when they build on a recognition that objects are always beyond the full grasp of concepts—as in the case of Morton's hyperobjects—concepts are still deployed above all as descriptive instruments, whose validity is measured by their fidelity to reality. Such fidelity is laudable but needs in my view to be supplemented by totalities understood as ideal types. My position here is not that we should choose between immersion in the micro-level and totalization, but that *both* approaches are needed. An exclusive adherence to either would blind us to how objects relate to the logic of capitalism. Choosing totalization alone would imprison us in idealistic models that disregard the contradictions generated by the impositions of reifying categories on objects. Equally one-sided would be to focus on objects without considering how these are shaped by the system. To use Morton's simile, the former perspective focuses too exclusively on the orchestra playing the tune of capitalism, missing how objects might dance in ways not dreamt of by the musicians. The latter perspective gazes at the dancers through a soundproof window, imagining the movements to be their autonomous creation. To study contradictions between the music and the dancing, we need both perspectives. If conceptual thought is one-sidedly rejected in favor of a conceptless immersion into the object, we will become blind to how the seemingly immediate is in fact mediated. Both concept and object are important, since it is the friction between them that provides the critical impulse.

The Entwinement of Society and Nature

Discussing the new approaches would be incomplete without mentioning the increasing interpenetration of social and natural realms, which constitutes part of the wider context in response to which these approaches have emerged.

Today it is a standard refrain in social theory that there is no longer any nature untouched by society (e.g., Castells 1996: 477f, Hardt & Negri 2000: 187, Jameson 1991: ix, Smith & O'Keefe 1980: 36). From a Frankfurt School perspective, it would be easy to view this as the outcome of the domination of nature. But at the same time, the idea of a domination of nature has come under fire for presupposing a rigid dichotomy between society and nature (e.g., Loftus 2012: 8f, Smith 2007, 2010: 45–8). This dichotomy is undermined not only by the increasing incorporation of natural processes into the circuits of capital but also by processes of naturalization through which the artificial increasingly takes on the semblance of a taken-for-granted natural environment (see Chapter 3; also Cassegård 2007, Vogel 2015).

In a variant of this argument, it has been suggested that the entwinement of natural and social processes today cannot be captured through the Frankfurt School's model of instrumental reason since it involves a new relationship to risk and unpredictability. Luigi Pellizzoni (2015) argues that today's neoliberal governing of nature differs from the old modernist striving to control nature by the fact that it makes the unpredictability of natural processes, markets, and technology a constitutive element in capital accumulation. Risk is not to be minimized or eliminated but managed or governed in profitable ways, as shown by examples such as the promotion of carbon markets and geo-engineering. Christoph Görg (2011) similarly argues that in today's post-Fordist capitalism, the domination of nature gives way to a qualitatively different goal, namely, the "reflexive mastery of nature" in which nature becomes a site for risk management as well as capital accumulation. Along similar lines, Jean-Pierre Dupuy (2013: 64ff, 75f) argues that the idea that technology has led to humanity's domination of nature is misguided. Technology today often aims at non-mastery, at creating things such as artificial life, genetic algorithms, robotics, and nanotechnologies that introduce unpredictability since they blur the life/death distinction.

These examples are claimed to show that technology is employed not to dominate nature in a conflictual fashion, but to promote and utilize natural processes and let production be guided by them, resulting in hybrids of social and natural elements. Such phenomena may be easier to grasp in terms of a production of nature, to use Neil Smith's (2010) term, than as a domination of nature.[4] Smith still stressed capitalism as the agent behind this production, but today many scholars are more inclined to attribute agency more generously to a variety of actors, or *actants*, or to the system formed by these actors as a whole. The idea of a conflictual relation between separate entities is then replaced by the image of an evolving whole coproduced by its constituent parts—an image that Moore seeks to capture with the term "web of life" and that new materialists usually refer to as an assemblage. Rather than a one-way domination by humanity of nature through instrumental reason, this coproduction is claimed to be characterized by a complex interrelation in which human attempts to control or steer nature is mediated through the unpredictable agency of non-human elements.

The entwinement of society and nature is real but is hardly a sufficient ground for claiming that domination has ended. As I argued in Chapter 3, the thesis of a domination of nature implies neither total control over nature nor any irreversibility of the process of increasing domination. Pointing out that nature too has agency or that human mastery over nature is faltering does not per se discredit the thesis. Contrary to Smith's argument that nature can't be dominated since it is produced, domination and production are not mutually exclusive. Domination doesn't preclude that nature is produced in the sense of being reshaped or fitted into a social form, which, as we have seen, is how Smith intends the production of nature to be understood. On the contrary, the imposition of form on matter is an essential part of capitalism's domination of nature since it takes place in order to appropriate the matter, the use-values, on which capitalism is dependent. It was to capture the potentially destabilizing role of the exchange of matter, as opposed to mere change of form, that Marx resorted to the idea of metabolism. When viewed as matter, rather than form, much nature is clearly not a product of capitalism. Many use-values provided by nature have been created over eons of time and cannot be conjured up by capitalism according to its wishes.

Furthermore, the idea of domination can perfectly well accommodate the idea that nature is not passive. As against Bennett's (2010: xf) claim that a nature that is vital can no longer be an object of mastery or domination in the same sense as when it is seen as passive or inert, vitality is hardly an obstacle to being dominated. After all, we speak of domination in relation to human beings, despite usually regarding humans as active agents. The mere existence of agency in an object does not prevent it from being dominated. It should give us pause that creating socio-natural hybrids or working through nature rather than against it is not a novel idea: relying on natural processes has always been central in agriculture, breeding, medicine, chemistry, and so on. Already Francis Bacon acknowledges the need to subordinate oneself to natural laws in order to be able to master nature. Hegel similarly points out that nature can be made to serve human purposes only by means of a "cunning" that allows humans to use the forces of nature against nature itself (Hegel 1983).

As both Pellizzoni and Görg point out, reflexive or neoliberal modes of governing nature represent a continuity with older modes of exploiting nature (Görg 2011: 59, Pellizzoni 2015: 67, 86). Mere reflexivity in the choice of means doesn't affect the capitalist system's overall goal of capital accumulation. The celebratory attitude to risk is still premised on an instrumental view of nature, making it disposable for intervention. This means that experiences of non-identity still have subversive potential. All systems that instrumentalize nature to some extent neglect non-identity, no matter how much they incorporate indeterminacy or the agency of objects in their mode of operation (Görg 2011: 59). In other words, even indirect modes of governing impose forms on what is governed.

We can, however, ask how useful the idea of a domination of nature is in a situation in which capitalism has shown itself to be amazingly unable to manage the environmental problems it has created, such as global warming. Is it reasonable to talk of domination in a situation when nature seems out of control? In addition to a theory of how nature is dominated, don't we also need a theory of how domination fails, inviting catastrophic consequences? To reply to this objection, we must first point out that the idea of domination doesn't presuppose that domination is always successful. On the contrary, the very point of materialism and of the form-matter dialectic is that domination and catastrophe go hand in hand, feeding each other. That catastrophes have become common is therefore not an argument against the thesis of domination. Already in the *Dialectic of Enlightenment*, the theory of domination was a theory of the modern enlightened world as the soil of catastrophes. However, to bring this catastrophic side of domination out better we need to go beyond Horkheimer and Adorno, who focused on the destructive consequences of domination for its victims, but had nothing to say about the possibility that the consequences might be destructive also for the system exercising the domination (see Biro 2011b: 235). I will therefore return to the idea of catastrophe in the next chapter, where I will bring the idea of dominating nature closer to Benjamin's and Adorno's idea of the permanent catastrophe.

Conclusion

This chapter has turned to Bennett and Morton, two thinkers who both seek to deny the duality between matter and ideas, nature and society. I have argued that while Bennett laudably highlights the independent agency of material flows and thereby pays respect to the primacy of the object, she tends to neglect the formation of second nature as a reified, non-vital reality that constrains and preforms the movement of matter. Morton is more attuned to the dark and subversive flavor of such reified objective structures, yet immersion reigns supreme in his writings too. Like Bennett, he affirms the idea of the non-identical. More than her, he mobilizes it in a critique of established concepts, above all "nature." Yet like her, he largely refrains from analyzing the structural background of the dominant position of these concepts in capitalist society.

The result is a position that, while having an environmentalist flavor, is also at odds with environmentalism as usually understood. While sharing the environmentalist critique of the Cartesian nature-society dualism for exacerbating human hubris over against non-human nature, Bennett and Morton appear unwilling to admit the validity of any environmentalist argument that presupposes a distinction between society and nature. Especially in Morton, conventional ecocriticism comes under fire for complicity in upholding the Cartesian dualism. There is thus a significant difference between environmentalist anti-dualism and the theoretically radical anti-dualism of these scholars. The problem with the radical position is a slippage: it dons an environmentalist costume by discouraging human hubris and encouraging solidarity with non-humans, but at the same time it pulls the rug out from under much environmentalism by disqualifying its arguments.

Against both, I have argued that we need not only to immerse ourselves in the pulsating world of matter, but also to distance ourselves from it to discern the workings of capital. We needn't shy away from the concept of capital, fearing it as inadequate to empirical facts. Firstly, we need to totalize to see through the real abstractions. Secondly, the point of totalities is not to represent empirical data but serve as heuristic tools. Thirdly, arguing for the value of totalization is not an argument against empirical research. A critical theory must pursue both. Totalizations must be accompanied by searching for what is non-identical in the objects and contradicts the concepts. Conversely, totalizations help us highlight the non-identical. It is by totalizing that we discover what doesn't fit in.

The concept of capital helps us see that the Cartesian dualism that Bennett and Morton criticize is a historical product, belonging to the real abstractions making up the negative totality of capitalism. Rather than declaring the duality null from the start, we should relativize the categories of society and nature in a way that is sensitive to historical context and that retains the tension between them. Solely focusing on vitality yields a picture that is overly sunny. Similarly, a sole focus on the slime of a decaying biosphere risks naturalizing this decay by failing to see it as a product of capitalism. As I argued in the previous chapter, criticizing the process whereby society and nature are reproduced by capitalism requires a methodology capable of running on two tracks: we need concepts adequate to mapping and grasping the predominant forms, but we must also sensitize ourselves to the complex interconnections and processes that are never entirely captured by these forms and that have the potential to destabilize them.

With the increasing capacity of society to intervene in nature—a capacity epitomized in the notion of the Anthropocene—it is crucial to diagnose the interpenetration of capitalism and nature correctly. This should be done without succumbing to a Cartesian dualism that absolutizes the dichotomy between the social and natural, but it should also be done without obliterating it prematurely, while leaving the system that generates it untouched. While criticizing the dualism, let us therefore also criticize the system at the same time!

10

Utopia, the Apocalypse, and Praxis

The discussions of anxiety and measure with which I opened this book point to the possibility of small incongruities leading to large transformations. The city of ever-increasing artificiality can tumble down and quality can be undermined and overturned by quantitative processes. Matter can subvert form. But what role does the subject play in these processes? What are the implications of a critical theory of nature for political action and for imagining alternatives to the present social order?

The typical procedure of a critical materialism is to mobilize the experience of non-identity for the purpose of critique, but merely criticizing seems insufficient if we also want to grope our way forward and create something better. Horkheimer and Adorno in particular have been criticized for a negativism that leads to a dead-end of despair.[1] As Jodi Dean points out: "If all we can do is evaluate, critique, or demystify the present, then what is it that we are hoping to accomplish?" (quoted in Bennett 2010: xv). This criticism can be extended to critical materialism as a whole. If the focus is on the immanent critique of a negative totality, what room is there for imagining alternatives? Isn't causal and practical materialism better equipped at providing guidance to political action, since they at least pinpoint relevant factors to take into consideration, such as the development of productive forces or the need to organize counter-struggles led by the proletariat?

In this final chapter I will argue that critical materialism goes excellently with both praxis and a utopian imagination. Critical materialism distances itself from practical materialism but doesn't reject praxis as such. Indeed, it frees up space for praxis by avoiding determinism as well as philosophies of history. A good example of the eminent compatibility of critical materialism with praxis is Marx himself, who didn't let the fact that his critique of political economy in *Capital* is a model of critical materialism prevent him from supporting the Paris Commune or leading the First International.

Below, I start by discussing what place that the conception of capitalism as a negative totality leaves to individuals as possible subjects of critique and political action. How can one conceive of a praxis that doesn't violate the primacy of the object? Isn't all action unavoidably an assertion of the subject over the object and hence to a certain extent idealistic, abstract, and dominating? There is, I argue, a straightforward answer to these questions. A critical praxis must visibilize the contradictions generated by the imposition of form on matter. This means that it can never presuppose itself as fully capable of mastering its environment. Such a critical praxis goes along well with political action as long as the latter is regarded as an occasion for encountering and perceiving non-identity.

Next, I argue that the many utopian visions in Frankfurt School writings are not an anomaly but part and parcel of negative dialectics. Their truth does not stem from correspondence to facts, but from their role in constellations where they negate other conceptual elements. They thereby become necessary precisely in order to do justice to the contradictory character of our present experiences. Finally, I suggest that utopian thought today must be understood in relation to the vision of an ongoing, unevenly progressing apocalypse that is gaining ground in contemporary environmentalism—a vision that I compare to Benjamin's and Adorno's idea of a permanent catastrophe.

The Critical Subject and Political Action

Criticizing the negative totality is impossible without conceiving of a subject capable of resisting or negating it. But what is the place of the subject in critical materialism? To what extent is the subject part of the totality it criticizes and from where does it gain its capacity to criticize? A useful model for thinking about the subject is offered by Adorno's remarks on constellations. To theorize the subject we should start, not with the idea of the system as a fully integrated totality, but in our experiences of non-identity between this system's categories and the reality to which they are applied. It is from these experiences—which

Adorno and Bloch exemplify with the vague sensation that "something's missing" or that "this cannot be all there is" (Adorno 1975: 368, Adorno & Bloch 1988:1ff)—that the critical subject is born. Non-identity is thus present from the start as its point of departure. When it reflects on itself, trying to be true to its experience, the subject takes shape as a constellation. It is neither a pure product of the system nor fully independent of it. The subject grasps itself through the juxtaposition of disparate elements, such as instincts, memories, dreams, and life plans. Instead of unifying them, the constellation brings to light the contradiction between the subject's conscious self-creation and the objective elements that disturb it. This means that the subject is not entirely subject, but always also object: I am always more than is subjectively meaningful to me. The fact that the subject can only ever be grasped as a constellation disqualifies all attempts to define it through a fixed, unitary identity. It is from this vantage point that Horkheimer and Adorno criticize the imposition of identity on inner nature in the *Dialectic of Enlightenment*.

The inner contradictions of the subject may prevent it from acting freely, but they also provide the critical impulses that generate political action. Understood as a mode of reflection on experience, negative dialectics can very well be accompanied by political action. As Vogel points out, "practice already contains within itself the moment of what Adorno calls nonidentity" (1996: 96, also see 2011: 199). Adorno himself admits as much when he writes that the "beneficial self-reflection of reason ... would be its transition to praxis" (Adorno 2005: 153). Not only are the critical impulses generated by negative dialectics impulses to act, but praxis in turn offers plentiful opportunities for apprehending non-identity.[2] Constellations go well with action since they shift in response to new experiences and therefore invite the subject to continuous self-reflection. They seem appropriate for a mode of thinking that "asks while walking," as the Zapatistas put it. They are devices less for passive contemplation than for thinking while acting, and while acting being true to the object and to the pain of a reified world. The fact that we do not have a philosophy of history to back up our action doesn't mean that we should be passive; only that we must think and reflect self-critically while acting.

Naturally, it is important not to overestimate the possibility of acting meaningfully in a situation characterized by the stubborn persistence of reified forms. To a large extent, the existent channels of political action are prepared by the system and fully integral to it. The environmental movement in particular has to a large extent become institutionalized in the form of green consumerism and a proliferation of governance networks and partnerships with authorities (Cassegård et al. 2017). That movements themselves have become caught up in existing forms and the identity-thinking associated with them goes a long way toward explaining our seeming "helplessness" in the face of climate change and thereby the so-called paradox of the Anthropocene: we know perfectly well that the system must be changed, yet seem unable to change it (Stoner & Melathopoulos 2015). In this situation, action to change the system is necessary but at the same time often appears almost impossible. Faced with this apparent impossibility, it is more important than ever to hold on to the experience of non-identity, as the source of what, despite all, may trigger a destabilization of the forms. Tragic and horrific as it might be, the fast pace of catastrophic events is at least a reminder of the fact that rapid, unexpected change is possible, and of the fact that the system, despite its appearance of changelessness, may in fact be as fragile as Bloch's artificial city and what Hegel calls quality.

Even as political mobilization gets going, it is important to remember that the collective subjects of movements are just as marked by non-identity as individual subjects. Movements encounter objective factors that disrupt them and force them to change direction. These factors are not only social but also natural. The exhaustion of natural resources may, for instance, mean that utopias of affluence cease to be viable. The progression of global warming may mean that even a socialist utopia will have to be built in a world where food is scarce, large swathes of land are uninhabitable, and coastlines are engulfed by acid oceans. The utopian strivings of the subject face not only the power of the system, but also the catastrophic reactions of objects to human intervention. At the same time, those reactions provide sources of critique of the system. The critical praxis that I propose emerges through an interplay between such subjective and objective elements. That movements are marked by non-identity is not only an obstacle but also a resource which may help them break out of the mold of preexisting forms and thereby endow them with the creativity and unruliness that will make them unpredictable not only to the system but also to themselves.

Experience and Communication

The question of critical theory's link to emancipatory political action necessitates a detour to the "linguistic turn" associated with Jürgen Habermas, portal figure of the Frankfurt School's so-called second generation. This turn led to the abandonment of the goal of a reconciliation with nature that was central to earlier critical theory. In its stead, pride of place was given to the ideal of non-coerced communication among human interlocutors (Honneth 1979). However, the price for this move is high from the point of view of a critical theory of nature. Despite Habermas's attempt to make room for an animal ethics in discourse ethics (Habermas 1993: 103–11), there is an inevitable anthropocentrism in the idea of basing the yardstick of critical theory on the normative presuppositions of human language-users and on the ideal of a domination-free consensus from which non-human life is excluded.[3]

Adorno's idea of reflection on experience can, I suggest, counterbalance the insensitivity to mute nature in Habermas's grounding of critical theory in language. This cannot be the place for a full elaboration of how an ethics of nature based on experiential reflection would look, but an important starting point is indicated by Jane Bennett when she states that the relevant public to address in political deliberation should not be exclusively human. "We need ... to devise new procedures, technologies, and regimes of perception that enable us to consult nonhumans more closely, or to listen and respond more carefully to their outbreaks" (Bennett 2010: 108). The willingness to let experiences of non-human nature shape one's thought and action is not entirely dissimilar from the willingness of language-users to try out propositions in communication with other people in order to find out to what extent they can withstand objections. Just as my position as a language-user compels me to listen to others if I want to claim validity for my beliefs, so my position as a living being, interacting with my environment, compels me to scrutinize my experiences, asking myself whether these experiences might inspire possible objections to the beliefs I want to uphold.

Nature, then, can be an interlocutor in my search for right beliefs and the right course of action, just as other human beings. This standpoint can accommodate the insights of Habermas's communicative ethics, since the experiences on which I must reflect include both experiences of nature and those of communicating with other human beings. Habermas is right that our opinions on truth and correctness can be validated only by communication, but communication should be extended to include *any* input from experience, including that of non-human nature. When I communicate with other human beings, what enables me to break out of monological self-centeredness is not what they say per se, but the fact that it helps me think new thoughts and gain new insights. This ability to make me think differently is not only possessed by human beings. For me to know for sure that a proposition is justified, it is not enough that all affected people would assent to it, as Habermas proposes. I *must* also know that no possible further experience would make me change my mind.[4]

Such a criterion of rightness does not mean reverting to a monological conception of reason, but is better understood as an attempt to expand a dialogical approach to the entire circle of my possible experiences, rather than limiting it to the narrow range of experiences I might have as an interlocutor in human communication. This would be a way to reorient critical theory back toward the ideal of a reconciliation with nature, but without giving up the ideal of autonomy. In this way Habermas's linguistic model can be reinterpreted to allow for critical input grounded in the experience of nature. This would not only help retain the notion of nature as the locus of utopian thought but also provide a safeguard against the anthropocentrism and the de-emphasis on materiality in Habermasian thought.

Taking on Second Nature

No matter how well suited a negative dialectics might be to action, the practical problem remains that a movement to stop the ongoing ecological destruction, which ranges from unprecedented species loss to runaway climate change, is up against overwhelming odds. The catastrophe is exacerbated by a fateful naturalization of society that makes the destruction appear inevitable. This is the background of the terror evoked by the notion of the Anthropocene: humanity is more powerful than ever, yet out of control.

Yet in this paradoxical situation it is possible to discern possibilities for a destabilization of capitalism's semblance of nature. The natural history that creates this semblance is also a history of resistance. The increasing capacity of society to intervene in nature turns the mode of intervention into a contested arena. Today, the "wilderness" cult (Guha & Martinez-Alier 1997) inspiring early conservationist movements finds itself sidelined and marginalized by two other environmentalist currents that both call for social interventions to regulate nature. On the one hand the adherents of the "eco-efficiency" current celebrate market mechanisms and technological fixes as the road to sustainability. Lauding the accelerating denaturalization of first nature, they enthusiastically buy into the naturalized lawlikeness of the capitalist market. A questioning of this second nature in turn comes about through a third current, the "environmentalism of the poor," which is driven by concerns about livelihood and brings attention to how social and environmental injustices are expressed and perpetuated in the naturalized operations of capitalism.

The struggles in which these environmentalisms are engaged emanate from the shifting relationships of natural history. They show, firstly, that environmentalism isn't necessarily oriented solely to saving first nature. The eco-efficiency camp reveals itself as a defense of *second* nature, since defending first nature is conditioned on its incorporation into the second nature of capitalist markets and technocratic governance, and the environmentalism of the poor mobilizes for an earth that is less a symbol of pristine wilderness than of the livelihood of common people. The second thing to note is that naturalization is not irreversible but gives rise to new waves of historization. Even as capitalism congeals into second nature, this second nature itself becomes the terrain of new struggles, and that means that the fateful conjunction of naturalization and ecological destruction isn't the last word.

To see what role these struggles play in relation to capitalism, we need to view them in the light of other conflictual relations that capitalism gives rise to. We have seen that clues to the multifaceted nature of the conflicts occasioned by capitalism can be found in how Marx's remarks on so-called primitive accumulation have been developed by scholars such as Nancy Fraser and Jason Moore. By showing how capital relies on both exploitation and expropriation/appropriation they demonstrate how its utilization of nature relates to a range of conflicts involving sexism and racism. These interact with class struggle in complex ways, and there is no guarantee that these struggles can be coordinated under a common front. Yet, as Fraser (2014: 72) points out, such a front might be possible if they come to understand themselves as struggles against capitalism.

The Reconciliation with Nature

The *Dialectic of Enlightenment* is often read as an expression of a loss of hope in political action, a gesture of despair whereby its authors lapsed into a totalizing critique that undercut its own foundations and led to political quietism.[5] What this narrative gets wrong, however, is firstly that it mistakes negativism for pessimism. As I have argued, the *Dialectic of Enlightenment* isn't propounding a pessimistic philosophy of history but aims at producing a critical impulse. Secondly, and most glaringly, the narrative fails to recognize that Frankfurt School writings not only before the publication of this work but also after it in fact abound with utopian visions and suggestive formulations about a possible post-capitalist reconciliation with nature.[6]

We have seen that Schmidt's *The Concept of Nature in Marx* fails to provide a clear vision of how the ruthless exploitation of nature might end. To start to fill in this gap, we should recall the many passages in critical theory that anticipate the utterly transformed, reconciled relation to nature that might be possible in a more rationally organized, post-capitalist society. In a striking passage, Adorno suggests that a free society—free from the compulsion to accumulate—might choose not to grow:

> Perhaps the true society will grow tired of development and, out of freedom, leave possibilities unused, instead of storming under a confused compulsion to the conquest of strange stars. A mankind which no longer knows want will begin to have an inkling of the delusory, futile nature of all the arrangements hitherto made in order to escape want, which used wealth to reproduce want on a larger scale.
>
> (Adorno 1978: 156f)

Providing a vivid counter-image to the predominance of instrumental reason he writes: "*Rien faire comme une bête*, lying on water and looking peacefully at the sky, 'being, nothing else, without any further definition and fulfilment' … None of the abstract concepts comes closer to fulfilled utopia than that of eternal peace" (Adorno 1978: 157).

Statements like these should be viewed in the light of the hope, expressed in the *Dialectic of Enlightenment*, for an enlightenment that no longer operates under the blind compulsion to dominate nature. To break out of this compulsion, reason must recognize its own roots in nature, thereby opening up for the mimetic relationship with nature that also serves as a model for the reflection on experience.[7] Envisioned here is that humanity, liberated from the spell of the value-law and the need for constant expansion, would no longer be compelled to ravage nature and might choose to limit its powers. A similar hope is conveyed in Benjamin's well-known formulation that the goal of technology should not be mastery over nature, but rather "mastery of the relation between humankind and nature" (1997b: 104).

Passages in the corpus of critical theory envisioning a similar reconciled relation to nature are not hard to find. One of the first to spring to mind is Marcuse, the Frankfurt School scholar who most emphatically embraced the notion of nature as a subject deserving of liberation, as expressed in the well-known proclamation that "nature too awaits the revolution" (1972: 74). This proclamation is made in the 1972 work *Counter-revolution and Revolt*, a book written in the wake of the student unrest and incipient new wave of environmental activism in the late 1960s that marks the climax of Marcuse's engagement with nature. Here he sketches the emergence of a new sensitivity harboring the possibility of a new way of interacting with nature, a new science, and a new technology.[8] In a statement that echoes Adorno's *rien faire comme une bête* he writes that the struggle with nature "may also subside and make room for peace, tranquility, fulfillment. In this case, not appropriation but rather its negation would be the nonexploitative relation: surrender, 'letting-be', acceptance" (1972: 69). It is significant that when Schmidt reevaluates his grim portrayal of socialism in the book on Feuerbach, he does so through a rapprochement with Marcuse, in whose writings he finds a resurrection of Feuerbach's idea of recognizing nature as subject in the moment of passive contemplation. The kinship between his and Marcuse's visions is evident in the emphasis Schmidt in his work on Feuerbach puts on the sensuous, contemplative interaction with nature as a gateway to apprehending it as a subject with its own *telos*, leaving it in peace (Schmidt 1973: 45, 47).

While mimesis and letting-be may be sufficient for individuals to experience moments of reconciliation with nature, the larger question is, of course, how society and the mode of production can be organized in a non-dominating way. As we recall, Schmidt's grim recognition in *The Concept of Nature in Marx* that socialism would not resolve the antagonistic relation to nature characteristic of capitalism has invited much criticism. As is clear from that work, however, his bleak view of the treatment of nature in the Soviet Union doesn't mean that he has lost hope in a better, more rationally organized socialist future in which nature would no longer be subject to blind domination, even if it would still have to be mastered (Schmidt 2014: 13). The problem with Schmidt's sweeping reference to socialism is that it fails to clarify how such a rational socialism would differ from Soviet-style technocracy and thereby makes it impossible to see what the former would consist in. That human beings must always work in order to survive, and therefore to some extent must treat nature instrumentally, doesn't mean that their relation to nature must necessarily take the form of egregious resource extraction, pollution, and habitat destruction.

Regardless of whether it is justified to label the Soviet system "socialist," any idea of socialism must build on allowing human beings to consciously direct the metabolism with nature and thereby ending the rule of value and the blind logic of accumulation. As Marx wrote, a socialist future would bring the metabolism with nature under the collective control of the associated producers, who would no longer be ruled by that metabolism as by a blind power (Marx 1991: 958f). Socialism therefore at least offers the *potential* of regulating the metabolism in an ecologically less destructive way than capitalism. What is envisioned here is a future in which the class divisions, the compulsion to accumulate, and the reified relations rooted in the logic of value would disappear. The problem with this vision, one might object, is that participation in the collective control is limited to human beings. What would prevent humanity from collectively ravaging nature in such a future?

Three answers are possible here, although they are all necessarily sketchy and incomplete. The first is a mere clarification: abolishing the rule of value is certainly no small step. If it leads to a suspension of the compulsive accumulation and expansion inherent in the logic of capital, then we will have stopped the great engine behind the increasingly catastrophic trajectory of capitalism's relation to nature over the last centuries, which has produced today's truly exceptional environmental destruction. Secondly, if the reifications rooted in the value law will disappear, then this will dereify relations to nature as well. Even though humans would still have to work with nature, a nature freed from reification would no longer be a mere resource to exploit. Here the possibility opens up for a sensitivity to nature that lets nature be a subject, as Marcuse puts it. The third answer is that nature will not let us rule it as we please. The catastrophic trajectory of capitalism may well lead to a depletion of resources of such magnitude that a future socialism would have very little left to plunder even if it would want to. If a socialism will ever come into being, it must happen through learning processes that help it appreciate what nature realistically has to offer and how humanity must behave in order to stop reproducing catastrophes.

Utopia as Critique

That Frankfurt School critical theorists have often endorsed utopian formulations is clear, but how do these go together with critical theory's generally suspicious view of affirmative visions? What theoretical status do these formulations have, if utopias are always unavoidably enmeshed in the realities against which they react?

The first answer is that utopias are determinate negations of the present. That means that they often function as tools for critique, rather than blueprints or programs. By startling us and producing shocks and shifts of perspective, they can free our imagination and help us think about what we want to accomplish. As suggested by Benjamin, utopias awaken the imagination from the mythic dream-world of capitalism. That is why utopias do not imply a break with critical thought even when they engage in wildly positive visions of the kind that Benjamin liked to illustrate with Fourier's dreams of lemonade seas, melting polar caps, and extra moons (Benjamin 1977b: 257). We should add that the critical effect of utopian visions is not produced by the vision alone, but, crucially, from its painful conjunction with a reality that is hostile to its realization. What Adorno refers to as the truth-content of art, its promise of happiness, is not its affirmative side, the sense of wholeness and perfection. The promise becomes visible in the disharmony created when art refuses to cover up the horrors of the modern world, thereby revealing the affirmative as illusory (Finlayson 2012). The same argument applies to the visions of reconciliation with nature, which acquire their strongest utopian luster in conjunction with the present reality of domination and destruction.

There is also a second, connected answer to how space can be created for utopian impulses within critical theory. Just as the seemingly positive—the utopia—functions as critique, so the seemingly negative—critique—has a positive side. Far from invalidating utopias, the fact that the "whole is the untrue" presupposes a utopian viewpoint from which the falseness of the negative totality becomes apparent: "The ray of light that reveals the whole to be untrue in all its moments is none other than utopia, the utopia of the whole truth, which is still to be realized" (Adorno 1993: 88). The confrontation of conceptual thought and object that from one side appears as immanent critique reveals itself from another side, as in a curious mirror, as an attempt to do the object justice. Another way to put this is that when we try to be true to our experience of the present, we mobilize thought in the form of constellations in which utopian visions have their place along with other conceptual elements that highlight the suffering, sadness, and injustice of the present.

This again illuminates the key role of constellations in the critical theory of nature. We have seen how they point to a solution to the Lukács problem, which arises when the status of natural science within critical theory becomes problematical through the rejection of causal materialism. In a similar fashion, constellations help us see the solution to the problem of how to conceive of the role of praxis and utopia in critical theory. In the dissonant whole indicated by the constellation, utopias and ideas for possible action are needed in order to articulate the longing rooted in the pain of the present. Instead of incorporating utopia into a conception of history as an evolving totality, as in practical materialism, critical materialism makes room for utopias as part of a critique of the present.

The Permanent Catastrophe

If utopian visions have a legitimate place in critical theory, what about their seeming opposite, the dystopian and apocalyptic visions that populate so much thinking about the future today? Like utopias, such dark visions can startle us and liberate our imagination—as Paul Ricoeur (1976: 25) put it, they can provide a "no-place" from which to view ourselves—meaning that they too are critical tools. To some extent, apocalyptic visions have had this function in modern environmentalism ever since the start of the latter in the early 1960s. Unlike other movements that mobilized people with the promise of a better future, environmentalism has often stood out by its "future-oriented pessimism" that mobilized by referring to doomsday scenarios (Thörn 1997: 322, 372).

That seemingly pessimistic visions can be a soil for utopias as well as ideologies can be illustrated by the debate around the Anthropocene. As mentioned in the introduction, this is a paradoxical notion that despite connoting unprecedented human power over nature is apt to inspire feelings of powerlessness. Several studies criticize this concept in favor of that of the "Capitalocene" (Malm 2016, Moore 2016a). The argument is that the former is ideological since it masks the fact that it's not humanity in the abstract but capitalism that bears the responsibility for the unfolding ecological catastrophe. This argument is correct, but another way to criticize this problematic concept is by showing that it contains a utopian kernel that can be discovered by immanent criticism, namely, the unfulfilled promise of a liberated humanity able to reshape its world through an interplay with nature. Fourier's extravagant visions can be seen as an expression of this kernel (Cunha 2015). In an age in which geo-engineering threatens to become capitalism's next large-scale step toward integrating itself fully into the circuits of nature, such utopias can, of course, be criticized as outrageous. Against such technocratic hubris it is easy to nod assent to the sober posthumanist recognition of the limits of human agency and endorse the hope that such recognition may become the basis of a new, humble, and caring relationship to our environment (Bennett 2010: ixf). Critics of posthumanism have a point when they assert that humanity today has greater capacity than ever to affect the earth (Hamilton 2017). But the critics miss that this capacity is itself out of control: rather than constituting real mastery, in the sense of increasing our freedom, it confronts us as an incapacitating second nature and is therefore experienced as powerlessness. The capacity to affect the earth resides in a system that with blind automatism pursues value even at the cost of squandering huge amounts of use-value. This is the background of the peculiar mixture of high awareness and ideology embodied in the notion of Anthropocene: the age in which humanity itself has become a terrifying natural force, threatening both itself and other species with extinction. Posthumanism correctly apprehends this powerlessness but naturalizes it, missing its structural roots in capitalism. To be convincing, its ethics of care needs to be complemented with a political struggle to end the rule of capital and value.

Catastrophe, the Primacy of the Object, and the Critical Subject

Catastrophes are the primacy of the object writ large. This helps explain why visions of the present world as apocalyptic are at least as prevalent in critical theory as utopias. Unlike history as progress, history as permanent catastrophe cannot be understood as the result of subjective intentions. In today's ecological crisis, the meaningfulness of history is again broken up, just as in the interwar years when Benjamin and Adorno developed the idea of permanent catastrophe. Benjamin's iconic image of the angel of history being blown into the future from a rising pile of debris portrays historical progress as an escalating catastrophe (Benjamin 1977b: 255). As both he and Adorno argued, hope does not reside in further progress, but in the possibility of interrupting the catastrophe—by pulling the emergency brake, as Benjamin expressed it (1996: 402). But how can we pull this brake, and how do we distinguish the catastrophe that will end capitalism from the catastrophes that constitute its normal run?

In recent years, a significant transformation of the apocalyptic motif appears to be under way in environmentalism. Rather than being invoked as a vague future threat, the catastrophe is often portrayed as ongoing, as already having started. As Eddie Yuen puts it, paraphrasing William Gibson: "The catastrophe is already here, it's just not evenly distributed" (Lilley et al. 2012: 130 n2). A rhetoric of ongoing or inevitable catastrophe characterizes groups as different as the Dark Mountain collective in the UK, the *collapsologie* network in France, the Tribunal for the Rights of Nature that seeks justice for injuries perpetrated on Mother Earth, and a variety of justice movements that protest against the racist, classist, and sexist distribution of loss. Hegel's discussion of qualitative change throws some light on this phenomenon. The accumulated weight of extinctions, crop failures, pandemics, and similar disasters today can result in a shift of perspective similar to what he describes in connection with the sorites paradox. A qualitative shift occurs when we suddenly see the catastrophe as prefigured in the first grain of sand removed from the heap. Similarly, the series of environmental catastrophes today may result in the entire history of what we've been taught to view as progress suddenly becoming visible to us as an incipient catastrophe. Seen from that perspective, it is not now that domination has suddenly started to falter but now that the catastrophic nature of the entire system becomes clear. The vision of history as permanent catastrophe results from such a shift. However, Hegel's account is still idealist in portraying the qualitative shift as unrelated to praxis. It doesn't link the surprise at the shift to any action on the part of the subject. But we can, of course, act when we realize that the heap is disappearing. We can try to prevent the disappearance or turn to another activity since continuing to remove grains of sand stops being meaningful. What characterizes the normal run of catastrophes is that they seemingly unfold without the subject's active participation. They are played out between capitalism and nature, without the subject being fully aware of them or at most with the subject as a contemplative onlooker. The shift to a view of history as permanent catastrophe, by contrast, involves a questioning of the contemplative framework. In that questioning a move to praxis, to reaching for the emergency brake, becomes possible. That possibility remains even in the darkest visions of contemporary environmentalism.

I have stated that a critique of the present order must be carried out through an interplay between subject and object. What matters is not to rely on the objective catastrophes alone, but to use them for critique. This raises the question of how to understand the relationship between the critical subject and non-human nature. In the *Dialectic of Enlightenment*, Horkheimer and Adorno portray the subject's recognition of itself as nature as linked to a relaxation of the drive for self-preservation, a way of breaking out of the self-perpetuating cycle of domination that turns society itself into a new nature. The most vivid illustration of such a recognition is perhaps when Adorno describes the mimetic identification with a wounded animal, in which the human subject reconnects to its own suppressed animal side while the animal is acknowledged as a subject. Unfortunately, there is a lack of concretion concerning how such a compassionate realization would be linked to praxis, although Adorno elsewhere writes about the redeeming qualities of the spontaneous impulse to act that, impatient with argumentation, refuses to let the horror go on (Adorno 1975: 281).[9]

With the increasing frequency of catastrophes, a different way for humanity to recognize itself as nature is revealed than the one Adorno describes, namely, through the insight that we too are vulnerable objects at the mercy of overwhelmingly powerful forces. This insight is not necessarily linked to a relaxation of the drive for self-preservation. It may be linked to the opposite, a renewed combative spirit. Where it brings about a suppression of the mimetic impulse, it can lead to a barbaric development in which, as Horkheimer warns, self-preservative reason brutally justifies itself as natural, as in Social Darwinism or fascism (Horkheimer 2013: 116–27). This possibility fuels the visions of an exterminist future or a Hobbesian war that are popular in postapocalyptic fiction as well as in social scientific scenarios (e.g., Frase 2016, Urry 2011). Where, however, the mimetic impulse is instead preserved, self-preservation can take the form of a solidarity with non-human nature against the system that inflicts the suffering. Such a recognition of oneself as nature needn't be conceived of merely in terms of compassion, as in the mimetic impulse arising at the sight of a wounded animal. The moment we combine the compassion with a readiness to act, the mimetic impulse can also turn into a revolt against the cause of suffering.

Clearly, the latter option is where a critical theory of nature would like us to go. But care needs to be taken lest the solidarity with non-human nature becomes an excuse for the abdication from subjecthood. This fantasy is expressed in the identification with junkspace, described by Rem Koolhaas (2002) as the corrosive language of matter—diminishing wattage, cracks, and malls or nightclubs that imperceptibly turn into slums. Faint echoes of this fantasy can be heard in autonomist ideas that celebrate the general corruption eating away at Empire, the desertion that will drain the energies away from the system (Hardt & Negri 2000). In its extreme variant this is a trust in automatism, in the idea that we can renounce our position as subjects and simply fall in line with the general decay. Fredric Jameson acknowledges junkspace as a subversive force, "a virus that spreads and proliferates throughout the macrocosm," but points out that it is also a space of commodification that, in the absence of utopias that negate it, is harmless to capitalism (Jameson 2003: 73). It's fine to identify with and ally with the junk, the viruses, the toxins, the shunned and abject, but it cannot absolve us from the need to be subjects also. We can act in tandem with the catastrophes and interplay with them—but trusting in them blindly is no better than trusting in capitalism.

How, then, should we relate to catastrophes if we want to criticize the system? We know several dead ends to be avoided. The first is to conceive of the subject as nothing but subject. Horkheimer and Adorno correctly identify this as idealist hubris: the only way out of destructive compulsion is to recognize oneself as nature. A second mistake is to passively trust in the catastrophes alone to undermine the system. This is likely to lead to barbarism rather than socialism. A third dead end is to abdicate from subjecthood by identifying with corrosive objective forces. Here the subject doesn't wait passively for the catastrophe to play itself out and instead actively tries to become part of it, but in both cases the interplay between subject and object disappears. A fourth and even worse option is to identify with nature in such a way as to justify a continued emphasis on ruthless self-preservation.

There is one option left, however—that of a joint struggle with nature against capitalism. Here the solidarity born in the moment of mimetic identification with nature is preserved while the drive to self-preservation is channeled against the system. Perhaps the slogan "We are nature defending itself" that could be heard during the weeks of the climate summit in Paris 2015 expresses this solidarity. Based on such solidarity, a praxis could take shape that unfolds as an interplay between subject and object, without relying overly on either. The critique of reified forms must be carried out *by* the subject but *with the help* of matter, which implies constantly watching out for, interrogating into, and responding to catastrophes—the clearest language through which non-identity is communicated to the subject. This praxis would make use of constellations as its tool for orientation and self-reflection in order to bring out experienced non-identity as clearly as possible.

What, more specifically, is the role of catastrophes in this joint struggle? Firstly, they give the lie to the idealism of capitalism, its presumption to adequately capture reality through its categories. What every catastrophe surely reveals is the hollowness of that presumption and the discrepancy between what most people find valuable and what capitalism treats as value. Secondly, catastrophes prevent socialist movements from becoming idealist. As mentioned, a future socialist society must adapt to what nature will allow it to do. Global warming and the exhaustion of natural resources ensure that it cannot rely on the wealth-generating productive apparatus of capitalism to fall into its hands like a ripe fruit. The rosy picture of history as an escalator of ever-increasing material prosperity must be discarded.

Whenever catastrophes come, mourning may well be necessary (Weber Nicholsen 2002). But mourning by no means implies quietism or despair (Cassegård & Thörn 2018: 569ff). Even if the catastrophes are already here, there is much left to do in the way of redressing wrongs, helping victims, and salvaging and redistributing what is still "worth a damn," to quote Evan Calder Williams (2011: 8). And, if we don't let the notion of catastrophe itself take on the function of a paralyzing ideology, there will also be many struggles left to fight—not least to counteract value as a medium for relating to nature and other people.

Conclusion

This book has made the case for a critical theory of nature built on the critical materialism developed in the Frankfurt School. Such a theory takes aim at the forms through which capitalist society regulates its interactions with nature. It criticizes those forms by highlighting the contradictions, or non-identity, between them and the matter on which they are imposed. In the course of the book I have attempted to clarify the outline of this theory, while also suggesting how it needs to be developed in order to overcome a number of problems. I have shown how Adorno's idea of constellations is crucial to solving what I have called the problem of the outside and the Lukács problem. I have also devoted attention to the problem of how to delimit and clarify the boundaries of the criticized totality in such a way that it best captures the contradictory relation between capitalism and nature.

In this final chapter, I have turned to the last remaining of the problems raised earlier in the book, namely, the problem of utopia and the possibility of a post-capitalist reconciliation with nature. In contrast to Bloch's unabashed embrace of utopian hope, Adorno and other members of the Frankfurt School easily come forward as overly negativistic and one-sidedly critical. This negativism has invited the criticism of stymying political energies. But as I have argued, negative dialectics is not purely negative. Frankfurt School writings abound with utopian visions and attempts to theorize reconciliation with nature.

To make sense of these, we should free ourselves from the commonplace depiction of Adorno, Horkheimer, and their colleagues as aporetically caught up in a totalizing critique that undercuts its own foundations. That depiction overlooks that what drives their criticism *is* a utopia, that of a world in which the non-identical, in ourselves as well as in nature, is no longer oppressed by the systems that dominate the earth today. Their writings may be dark, but they are not devoid of stars.

The idea of constellations shows why utopias have an important and legitimate place in the critical theory of nature. The emphasis on criticizing the present doesn't mean that critical theory can do without a utopian imagination. Utopias are neither escapist fantasies nor realistic blueprints, but indispensable critical instruments and subversive fantasies that help us question the status quo and imagine alternatives. Critical theory stands out not only by its emphasis on criticizing ideology, but also by holding onto a utopianism that is lacking or more subdued in many other Marxist currents—a utopianism that finds nourishment in the progression of catastrophes around us. It shouldn't be forgotten that these utopian anticipations appear utopian because they are antithetical to the real, capitalist society that today blocks their realization. It is to fight this society—which has always exploited real nature for the purpose of capital accumulation and denied the human capacity to act by presenting social processes as natural—that their unfulfilled utopian promise should be kept in mind.

Notes

Chapter 1

1. Critical theory insists on the resistance of matter to form. In this sense it is materialistic while Hegel is not. Hegel (1969: 450) states that matter is only visible to us as determinate, that is, in conjunction with form. In contrast to that, what I call critical materialism holds that we can perceive the effects of matter in the subversion of form. This is, I believe, most convincingly argued through an appeal to experience. When a form is imposed on matter, for example, the value-form of a commodity, we are still able to see it as more than a commodity. In other words, we can apprehend it more concretely than the perspective of a single form allows. See Chapter 3 and see Cook (2011: 40f) for a discussion of the latter idea in connection with Hegel's form–matter distinction.
2. Although this book focuses on the forms constitutive of the capitalist economy, I readily admit that forms constitutive of the modern state as well as of gender relations are equally important to criticize. For an attempt at theorizing how climate change might affect political forms, see Wainwright and Mann (2018).
3. See Testa (2007) for a history of the concept.
4. Examples of authors who understand reification as viewing humans or human processes as things: Berman (1989: 141), Feenberg (2014: 62), Hailwood (2012: 884ff), Williams (1983: 35), among others. Honneth (2012) too belongs to this camp since he redefines it as forgetfulness of recognition. The basis for his interpretation is Lukács's reference to Marx's description of commodity fetishism as a phenomenon that makes social relations appear as a relation of things. It seems to me, however, that this overlooks the fact that Lukács applies the concepts of reification and thing-form to a much wider set of phenomena, not least by describing them as characteristic of the "bourgeois thinking" characterizing idealist philosophy, which implies that they are applicable to the world as a whole, not just to humanity.
5. For some examples, see Bell (2017), Biro (2011a), Buck (2013), Cook (2011), Feenberg (2014), Flodin (2011), Hall (2011), Jarvis (2004), Luke (1994), Malm (2018), Pellizzoni (2015), Pensky (2005), Sanbonmatsu (2011a), Stoner &

Malthopoulos (2015), Truskolasky (2014), Vogel (1996), Westerman (2019), Wilding (2008), to name a few.

6   I concur with Wilding (2008: 48) that it is imperative to analyze the causes of the catastrophic environmental crisis and that "[r]esources for such a theory can be found in the ideas of the Frankfurt School." However, the emphasis of the critical theory of nature that I am proposing in this book differs from his in that my approach is closer to Adorno's and that I find crucial resources in the latter's materialism, his idea of natural history, and his use of constellations.

7   For catastrophe as permanent or continuous, see e.g. Adorno (1978: 192, 1973a: 320), Benjamin (1977b: 255, 1999: 473).

8   On the concept of nature, see Soper (1995), Williams (1980), Worster (1994), Smith (2010).

9   Recent decades have seen a proliferation of such approaches, including the pioneering early work of eco-socialists like James O'Conner, Ted Benton and Michel Löwy, literary and philosophical scholars like Raymond Williams and Kate Soper, environmental sociologists, historians and economists like Stephen Bunker and Joan Martinez-Alier, the treadmill of production approach of Allan Schnaiberg, radical geographers like David Harvey and Neil Smith, the contemporary eco-Marxism represented by John Bellamy Foster and Paul Burkett, the human ecology of Alf Hornborg and Andreas Malm, the world-ecological approach of Jason Moore, and many more.

10 Smith qualifies this constructivism by stressing that his position shouldn't be confused with the discursive constructivism fashionable since the 1980s (Smith 2006, 2010: 246) and that it doesn't imply that matter itself or natural laws like gravity are constructed or that producing nature means controlling it (Smith 2006, 2010: 87ff), or that economically irrelevant portions of nature like outer space are constructed (2010: 80).

11 This is a complex procedure which I will explain fuller in the book (especially Chapters 3 and 6).

12 E.g. Merchant (1990). For criticism of the dualism outside the Marxist camp, see e.g. Cronon (1992), Descola (2013). The above remarks on realism and constructivism have direct bearing on the debates around this dualism. Here is not the place to discuss this debate, which I will revisit in Chapter 8.

13 It is, of course, possible to object by pointing to the environmental destruction in the Soviet Union and China (for a reply to that, see Malm 2016: 277f). The main point is that even if these countries contained non-capitalist elements, they were still forced to accumulate by their place in the world-system due to economic and political rivalry with the capitalist world.

Chapter 2

1. From another angle a roughly similar tripartite division has been proposed by Elbe (2010: 12–29, 2013), although he doesn't focus on materialism. The term "critical materialism" is also used by Cook (2011) to designate Adorno's materialism.
2. As Creaven points out, "both Marx and Engels use the language of 'reflection' as a polemical device to distinguish their own position in the most general way from that of the Hegelian school … It has no other function in their writings. For whenever Marx or Engels go beyond general formulas … their analysis immediately becomes more subtle and qualified" (2000: 45).
3. To Hegel, causal relations are relations of externality, unlike those that he wants to lay bare, which are internal to notions (Hegel 1969: 446). See Fine (2001: 81–5, 96f) for a discussion of how Marx misconstrued Hegel in using the metaphor of inversion. To the extent that causality figures centrally in Hegel's dialectic, it is a teleological causality modeled on Aristotle that is quite different from the causality I associate with causal materialism (see Beiser 2005: 66f).
4. The latter work has fittingly been described as a "stormy marriage between Helmholtz and Hegel" (Rabinbach 1992: 81), Helmholtz being a German natural scientist. For a background to the texts in relation to the development of Engels's thinking, see Liedman (2018: 492–500).
5. Engels's use of the word "law" is telling—as Liedman (2018: 486) points out, Hegel never applied the word to dialectical processes.
6. See for instance Harvey (2010: 191), Liedman (2018: 616).
7. An example is the recurring discussion among Marxists about the relative weight of productive forces compared to relations of production (for a recent and vigorously argued example, see Malm 2016).
8. It is when this materialist safeguard is given up that the road opens up for the more unabashed celebration of contingency in post-Marxism, poststructuralism, and actor-network theory, which thus has an important ancestor in Althusser (see Söderberg 2017).
9. See Williams (1978) for a criticism of Timpanaro that highlights the fault line between causal and practical materialism. Contemporary defenders of the base-superstructure metaphor include, for example, Creaven (2000) from a critical realist perspective and Holt (2009) from the perspective of analytical Marxism.
10. For works questioning the centrality of causality to Marx, see Elson (2015), Gunn (1992), Harvey (2010: 193), and Ollman (2003: 27, 38).

11 Creaven (2000: 62f) correctly argues against treating the couplet of base and superstructure as the same as the distinction between being and consciousness. The former refers to vertical relations between structures while the latter refers to "the 'determination' of the consciousness of living individuals by the totality of material and social circumstances and relations."
12 An excellent discussion of the centrality of nature in these manuscripts is Saito (2017a).
13 For a further discussion, see Chapter 6.
14 For the development of the new Marx-reading, see Chapter 4.
15 It thus differs from idealist philosophies such as the Kantian system, which can very well be realist in the sense of recognizing the existence of objects beyond the grasp of thought, but refuse to grant these objects subversive potential. In refusing to posit "matter" as a foundational concept, critical materialism resembles what Toscano (2014) refers to as "materialism without matter."
16 Finlayson (2014) recognizes this but diagnoses it as an inconsistent vacillation. Against this, I want to stress that when Adorno calls for transcendent criticism, this is not transcendent in the ordinary sense (e.g., criticism that negates something abstractly based on an external yardstick). What is non-immanent is rather the experience of the object and its non-identity with the concepts. This may appear *logically* inconsistent but is true to the idea of negative dialectics.
17 As Moore (2015: 2) puts it: "Capitalism's governing conceit is that it may do with Nature as it pleases." Morton (2007: 91) similarly rejects the description of capitalism as materialistic and argues that in its disrespect for matter it is "much more like an idealism gone mad." Westerman is also clear on this point: only capitalism is so completely self-enclosed and self-validating so as to see nature "only as something to be transformed into commodities" (2019: 263). The idea of capitalism as a system of abstractions goes back to Marx (see e.g. 1993: 164).
18 On the difference between such critique and transhistorical theorizing, see also Schmidt (1981: 44) and Gunn (1992).
19 For instance, Schmidt stresses the importance of interpreting Marx's concepts as historically bound to or conditioned by capital (1976: 64–68). See Chapter 5 regarding the transhistorical application of Marxists concepts such as value and labor.
20 For these interpretations, see Adorno (1973a: 300, 1974: 22f, 1993: 20), Postone (1993: 75–80), Arthur (2002: 102f), and Banaji (2015: 34). For textual support, see Marx (1990: 255f). Bonefeld (2004) criticizes Postone's overemphasis on the idea of capital as subject for missing the negative character of that spirit,

that is, the contradictions. Rather than idealistically reproducing the categories of capital, one should note that Marx engages in an "ironical or parodical repetition" of Hegel, as Schmdit (1968b: 26f) points out.

21 On Adorno's concept of totality, see Heitmann (2018), O'Kane (2015, 2018a), and Rose (1976). Jay's (1984) treatment of this concept misses its negative aspects and its link to critique. For the idea of negative totality as read back into Marx, see Arthur (2002: 8) and Postone (2009: 72). Its link to critique is put well by Lange: "for Hegel, our task was to comprehend the logical, the natural, the scientific, the historical, the economic, the social and the psychological categories in their *final truth*, not – like Marx – in their *final falseness*" (2016: 266f).

22 Capitalism's "consolidation into a totality is itself irrationality, the totality of the negative," Adorno (1993: 87f) writes. Unlike the Hegelian totality the goal is thus to abolish it (Adorno 1974: 198, 1975: 348, Schmidt 1968a: 52, 57, 1976: 104ff, 1981: xvii, 1983: 19).

23 Fraser refers to them as capitalism's background conditions or back-stories (Fraser 2014, Fraser & Jaeggi 2018).

24 For a similar argument, see Schmidt (1981: 46, 52f). Another example is Arthur's argument about how preconditions of the system become "virtual" in the system itself (2002: 121, 124).

25 Bringing in natural science in this way is common in eco-Marxism. For further discussion corning this, see Chapter 7.

26 For instance, while Marx wrote a long discussion about precapitalist Germanic societies in *Capital* Vol. 3, it hardly belongs to the theoretical core of his theory.

Chapter 3

1 For animals, trees, and so on as a subject of moral concern in Adorno, see Sanbonmatsu (2011b), Mendieta (2011: 151), Flodin (2011), and Gunderson (2014). This chapter develops some ideas also presented in Cassegård (2017).

2 This ambivalence in the concept of nature also characterizes Benjamin's discussion on language and mute nature (1988).

3 Cook also uses the term "critical materialism" for Adorno's materialism in her excellent account (Cook 2011: 7–33). Other accounts of Adorno's materialism: Hall (2011), Jarvis (2004), and Truskolasky (2014). The longest sustained discussion about materialism in Adorno's own oeuvre is probably Adorno (1974).

4   Cook points out that Adorno combines the Hegelian idea that all is mediated with the Kantian idea of an object non-identical to its concept. Mediation—as in Hegel—concerns the realm of concepts, but concepts do not have the final word. This makes Adorno a materialist and prevents him from becoming a social constructivist (Cook 2011: 41).
5   This interpretation colors Habermas's influential account on Adorno. For critical comments on that, see Chapter 10 and O'Kane (2015: 190f, 2018a: 288).
6   As Rose (1976) points out, totality is a critical concept in Adorno. "Totalizing" isn't "total," and Adorno's account of present-day society doesn't imply that identity-thinking has triumphed entirely. This is also pointed out in Görg (2011: 57) and O'Kane (2018a: 289).
7   To put the object in relation to form-matter, Hegel writes that objects arise when form allies with content (Hegel 1975: 5f). Against this idea, the position I advance with the help of Adorno insists on non-identity, since our experience of objects is almost always more multifaceted that what can be captured through any specific form.
8   For Minamata disease, see George (2001) and Walker (2010).
9   Adorno (1975: 184–98); for discussions of this notion see Cook (2011: 9–15), Feenberg (2014: 159f), Hall (2011), and Jarvis (2004).
10 I will return to constellations in Chapter 6.
11 On the sensuous in Adorno's materialism, see Jarvis (2004: 80, 98), Schmidt (1983: 22), and Truskolaski (2014: 18ff).
12 Adorno uses the same argument against both Epicurus's and Soviet "dialectical" materialism, namely, that they both bring together the objectivity of physical nature with subjective perception through forms of reflection theory (in Epicurus's case the idea that matter emits "little images"; Adorno 1974: 212, 214, 1975: 207; cf. Truskolaski 2014: 15f).
13 A similar idea is expressed by Horkheimer (1988), who argues that the meaninglessness of matter means that materialism cannot be a mere inversion of idealism; that is, it cannot result in a metaphysical system of its own.
14 For this point in Adorno's critique, see Buck-Morss (1977: 56), and Hall (2011). For a partial defense of Lukács, see Westerman (2019: 4f, 241).
15 Adorno (1973a, 1975: 347–58); for discussions of this idea, see (Bell 2017), Buck-Morss (1977: 49–59), Cook (2011), Feenberg (2014: 162f), Pensky (2005), and Rose (1978, 38ff). For Benjamin's treatment of the dialectics of nature and history, see Benjamin (1985: 45ff, 91ff, 129ff, 165ff, 177ff, 223ff), Buck-Morss

(1991: 58–201), and Cassegard (2007: 37–42). Benjamin too employed the *chiasmus* to express the idea of natural history in his later writings: "No historical category without its natural substance, no natural category without its historical filtration" (Benjamin 1999: 864).

16 Although this is clearer in his earlier works (e.g., Lukács 1996: 62ff; see Cassegård 2007: 39f) than in *History and Class Consciousness* (see Westerman 2019 for how Lukács there overcomes his earlier romanticism).

17 See the interpretation by Buck-Morss (1977: 55ff).

18 For passages that support a "pessimistic" reading, see Horkheimer and Adorno (2002: 9, 13, 55, 185). At the same time, the authors criticize philosophies of history that impute meaning into history (e.g., Horkheimer & Adorno 2002: 185), a fact that indicates that their comments on the course of civilization can be read as polemical.

19 For examples of this criticism, see Habermas (1984: 373–86, 1992: 333, 1994: 106, 116–19, 126f), Schmidt (1986: 216f), Harvey (1996: 137ff).

20 As Adorno points out, the history of catastrophes is constituted by contingent events that "didn't have to be" (Adorno 1975: 315). Bonefeld writes that Adorno's argument about history as a progression from slingshot to atom bomb "does not say that the sling-shot contained the atom bomb as its necessary further development. The atom bomb is not the innate necessity of the sling-shot. Rather, it says that the atom bomb reveals the sling-shot, and as such, it reveals a history made universal as one of violence and destruction" (2014: 97 n22). Bell in a clarifying account argues that natural history by seeing history as constituted by a series of material events undermines idealist historiography (2017: 189).

21 In stressing contingency in Adorno's view of history I follow Allen (2017: 172). Adorno stresses that a turn to the better is possible at any time, even in the present; the point of criticism is to enable progress in this sense, which has hitherto not been realized (Adorno 2005: 150).

22 For the idea of natural history as a critical tool, see Buck-Morss (1977: 49, 1991: 59) and Cook (2011: 2).

23 This point is made well in Buck-Morss (e.g., 1977: 58) and Gunster (2011: 220f).

24 Dark Mountain is a group of writers, artists, and activists who came together in 2009 to put words on the experiences of loss accompanying the decay and collapse of civilization (see Kingsnorth & Hine 2017).

25 For criticisms of the wilderness idea, see e.g. Cronon (1995), Guha (1999), and Radkau (2008).

26 For instance, Adorno's idea of natural history can be used to criticize the idolization of pure or pristine nature in Deep Ecology (Bell 2017: 191, Buck 2005).
27 For naturalization, see Cassegård (2007). This phenomenon is also described in Vogel (2015), Smith (2006: xiv), and Harvey (1996: 186).
28 Moore argues that "'environments' are not only fields and forests; they are homes, factories, office towers, airports" (2015: 45f). For urban environmentalism, see Anguelovski and Martínez Alier (2014), Heynen et al. (eds.) (2006), Loftus (2012), Swyngedouw (2015).
29 For criticisms of the anthropological/transhistorical focus on the *Dialectic of Enlightenment*, see also Biro (2011b: 235), Breuer (1993: 273), Postone (1993: 119), Smith (2007: 24, 2010: 46), and Smith and O'Keefe (1980: 33). Others defend the transhistorical scope of the argument (Pellizzoni 2015: 86, Wilke 2016).
30 It is well documented that Adorno and Horkheimer excised references to class struggle from the text, partly out of political self-censorship (van Reijen & Bransen 2002).
31 As environmental history shows, ecological degradation starts long before capitalism (e.g., Radkau 2008).
32 In the *Dialectic of Enlightenment*, class is referred to in the analysis of fascism (Adorno & Horkheimer 1997: 172). References to capitalism continued to be important to Adorno, who, unlike Horkheimer, never wholly succumbed to the idea that oppositional forces had become integrated and neutralized in "administered society" (Breuer 1993: 268). For examples, see Adorno (1987, 2005: 159, 344 n11, 2018). See O'Kane (2018a) for a discussion of capitalism in Adorno's writings from the 1960s that helped inspire the new Marx reading.
33 See e.g. Breuer (1993) and Buck (2005).

Chapter 4

1 Part of this chapter builds on and expands arguments in Cassegård (2017).
2 Apart from the eco-Marxist criticism discussed in this chapter, other who have criticized Schmidt include, for example, Neil Smith (see Chapter 8), Holt (2009: xi, 24ff, 65), and Saito (2017a: 64f, 79, 84f, 91, 97).
3 "Metabolism" in Marx's writings stands for the material side of the relation between man and nature, the labor process as a process of nature. Marx used it to point to how the separation of town and country disrupts the circulation of

nutrients, leading to soil degradation. See Marx (1990: 198, 283, 287, 290, 1991: 878, 949, 1993: 410), Foster (2000), Schmidt (2014: 76–93), and Saito (2017a: 63–137).

4 For a clear statement of Schmidt's criticism of Lukács and other Western Marxists for neglecting nature's otherness and reducing Marxism to a philosophy of history, see Kocyba (2018).

5 For Adorno's influence on Schmidt, see also Cook (2011: 23f, 26, 28f).

6 Here Schmidt's position agrees with those who argue that alienation from nature cannot be wholly overcome (e.g., Biro 2000, Hailwood 2012).

7 In his criticism of Schmidt, Saito (2017a: 84f) misunderstands this point. Saito sees Schmidt's negative ontology as a reflection of Jacob Moleschott's and Feuerbach's transhistorical ontology and implies that a disagreement exists between Schmidt and Adorno since the latter refused to recognize the non-transcendability of natural laws. What Saito overlooks is that Schmidt's negative ontology is not a reference to transhistorical ontology but to Adorno's non-identical object. There is no incompatibility between Schmidt's and Adorno's standpoints on this point. Saito also conveniently omits to mention a fact that sits ill with his interpretation, namely, that Schmidt was still *critical* of Feuerbach's transhistorical ontology in the *Concept of Nature in Marx* (unlike in later works like *Emanzipatorische Sittlichkeit*).

8 Adorno anticipated several themes in the new reading (see Adorno 2018). In viewing the economy as a negative totality to be subjected to materialistic critique, the new reading can be described as negative dialectics as applied to the political economy. On the roots of the new Marx-reading in the Frankfurt School, see Bellofiore and Redolfi Riva (2015, 2018), Bonefeld (2014: 4), Elbe (2010, 2018), O'Kane (2015, 2018a), and Reichelt (1982). For Schmidt's role in developing this reading, see Kocyba (2018). Many assumptions of this reading are shared with the broader current of value-form theory and open Marxism (e.g., Arthur 2002, Bonefeld 2014, Postone 1993).

9 That capitalism forms a negative totality is underscored by Schmidt already in 1968. "There is in fact in Marx something like what we might call an ironical or parodical repetition of certain categorical transitions in the Hegelian logic," for example, when he says that value "moves." For instance, when Marx claims that living human beings are treated in *Capital* as the personification of economic categories, then this sounds idealistic, but that is because the world forces idealism on people (Schmidt 1968b: 26f).

10 He thereby avoids the weakness—pointed out by Arthur (2003: 132f), Bonefeld (2014: 6, 10f, 79), Elson (2015: 144), Kincaid (2005), McNally (2004), Ollman (2003: 183f), Saito (2017a: 15f, 100, 113–19)—that the new Marx-reading and/or value-form theory sometimes overemphasize the social logic of forms while neglecting antagonisms, contradictions, and the material world.

Chapter 5

1 Since I am interested in showing how the metabolism with nature is conditioned by the rule of the value-form, this is a value-form approach to nature. A pioneering work in this approach is Burkett (1999). As I discuss in Chapter 8, Moore (2015) develops this approach while moving considerably beyond Marx. For an analysis that, in opposition to pure value-form theory, stresses the need to focus on use-value, see Altvater (1993). For an excellent collection of contributions on value, see Elson (2015). There is now a sprawling debate regarding the relation between value and nature, spanning both more orthodox and more unorthodox positions. See for instance the special issue "Value in Capitalist Natures" (ed. Kenney-Lazar & Kay) in *Capitalism Nature Socialism* 28(1), 2017. A wide variety of non-Marxist approaches to value also exist in the field of environmental history and economics; what is common to the latter is their rejection of the labor theory of value and their attempt to see nature as a creator of values (e.g., Cronon 1992: 148–51). For a critique of some of these approaches, see Foster (2018), Foster and Burkett (2016, 2018).

2 For the retrospectivity involved in the determination of value and abstract labor, see Rubin (1973), Heinrich (2012: 53), and Burkett (2014: 58).

3 Marx (1990: 163–70). On the origin of fetishism, see also Marx (1991: 267f, 965–70).

4 For the concept of real abstraction, see also Engster and Schlaudt (2018), Reichelt (2007), and Toscano (2008).

5 The argument that fetishism obscures the role of nature is also made by, for example, Altvater (1993: 183), Huber (2017: 45), Schmidt (2014: 68), and Yates (2011: 241).

6 Connected to this is the fact that nature, unlike labor, comes forward as a "free gift" to capital (see Burkett 1999, 2014: 69–78, Marx 1992: 431f, 1993: 765). Even where nature is priced, that is, turned into private property, its use no longer reflects the dynamics of value proper but rather that of rent (see below).

7   Where Marx addresses the question about the difference between human and non-human labor (in the argument about human beings being able to plan their activities while bees and spiders are not, Marx 1981: 68, 1990: 284), he does so in general, transhistorical terms that lack connection to capitalist value-creation. This distinction, which goes back to Marx's statements in the Paris Manuscripts, certainly implies a certain anthropocentrism, firstly by drawing a boundary between humanity and the rest of nature, and secondly since his concern in these manuscripts is mainly with humanizing nature to suit human needs (see Benton 2011). It is, however, possible to provide a different, non-anthropocentric answer by making inferences from his theory of value in *Capital*, which is what I try to do here.

8   This proposal originates with Sergei Podolinsky. For a criticism of it from a different perspective than mine, see Foster and Burkett (2016).

9   The thought experiment rests on several simplifying assumptions. I regard demand for the commodity as inflexible and simply assume that it will be sold. I also assume that wages are inflexible (so that wage-labor isn't increased, for instance, by moving production to countries with cheaper wages). I also disregard the possibility that increasing the quantity of workers can have the effect of increasing productivity (e.g., through a division of labor). Finally, I disregard that technology has costs that in some instances (e.g., start-up phases) may off-set the tendency for technology to cheapen production.

10 I therefore disagree with Walker (2017), who argues for a unified measure of labor-nature time since a surplus must be created regardless of whether the work is performed by human or non-human labor. "That ploughman's horse had better produce more in grain than it consumes or it will be put out to pasture; and the same is true of the wage worker" (2017: 55). However, the difference between the contributions of wage-labor and "nature" becomes visible when one varies their proportion. To use his example, when horses are introduced in agriculture (assuming this agriculture to be capitalist), they increase productivity and hence *lower* the value of the grain, a phenomenon that doesn't happen if the increase is in human labor instead of in horses.

11 It doesn't hold for most forms of slavery, where the slave-owner provides the costs for the reproduction of the labor force. This means that slavery approximates modern wage-labor better than serfdom since serfs are to a greater extent responsible for their own subsistence.

12 This argument opens up for further exploration of use-value as a complement to the analysis of the value-form. Postone has been criticized for not pursuing this option (Bonefeld 2004: 115). For a defense of Postone that stresses the potential for such development, see O'Kane (2018b).

13 This point is forcefully made by McNally (2004) who argues that Postone fails to see that the contradiction between abstract and concrete labor is central to Marx's critique. Arthur too points out that such binaries perform a "critique of identity," subverting what from the point of view of value appears like identity (2015: 73).

14 For clear statements of this argument, see Moore (2014a: 17, 2015). As Marcuse says, "nature appears as that which capitalism has *made* out of nature: matter, raw material" (1972: 62). Sohn-Rethel too points out that only with the rise of commodity exchange does a scientific concept of nature emerge, sharply distinct from magical and mythology (1978: 55).

15 In the debates concerning the transhistorical applicability of Marx's concepts, some scholars argue for a transhistorical reading of concepts such as value and abstract labor (e.g., Elson 2015, Hussain 2015, Saito 2017a) whereas others reject such a reading (Arthur 2003, Bonefeld 2014, Postone 1993, Rubin 1973). Some scholars, eager to prevent a transhistorical reading of Marx's theory, argue that even concepts such as concrete labor and use-value are bound to capital (e.g., Trenkle 2014, Yates 2018). The position I support views value as typical of capital while use-value is not necessarily bound to capital (see also e.g. Postone 1993: 360, Gandler 2018, Burkett 1999).

Chapter 6

1 For the Marcuse-Habermas debate concerning this, see Feenberg (1995, 2014: 194–200), Alford (1985), Habermas (1971), Marcuse (2002: 157–61).

2 For criticism against the Frankfurt School's failure to accommodate natural science, see also Görg (2011: 55), Wiggershaus (1994: 604, 609).

3 For discussions of the Lukács problem, see Foster (2000: vii), Feenberg (2014: 121–49), Jay (1984: 116), Loftus (2012: 62–6), Pellizzoni (2015: 193–6), Vogel (1996: 13–27), Westerman (2019: 241–74).

4 Lukács writes that

> the dialectics of nature can never become anything more exalted than a dialectics of movement witnessed by the detached observer, as the subject

cannot be integrated into the dialectical process ... From this we deduce the necessity of separating the merely objective dialectics of nature from those of society. For in the dialectics the subject is included in the reciprocal relationship in which theory and practice become dialectical with reference to one another.

(1971: 207)

Some commentators have seized on this passage to argue that Lukács acknowledges the possibility of a dialectics of nature (Foster 2016a: 412f, Foster et al. 2010: 219, Rees 1998: 245). This must be weighed against the fact that dialectics elsewhere in *History and Class Consciousness* is linked to praxis and described as essentially taking place between subject and object. As Vogel points out, Lukács leaves the assertion that nature is dialectical "entirely without foundation" since he fails to provide any clue as to how such a dialectics could be known (1996: 19). Regardless of that, the crucial point is that Lukács insists that the two kinds of dialectics are different and hence cannot provide the foundation for a unity of method.

5 These propositions fit logically together if interpreted as implying that Lukács was a social constructivist regarding the category of nature but a realist concerning nature as a material reality (Feenberg 2014: 130, Westerman 2019: 242f; also Lukács 2000: 97). Yet neither proposition contributes to elucidating the dilemma of how dialectics is supposed to be a critique of reification if it must recognize reified thinking as legitimate in relation to nature.

6 Foster (2016a) makes a tendentious reading of the *Defence*, suggesting that Lukács here endorses a purely objectivistic approach in line with Engels's dialectics of nature. Foster's reading ignores Lukács's repeated insistence that nature can only be apprehended by subjects and is therefore necessarily mediated by subjective categories. As Foster's own reference to Hegel's doctrines of Being and Essence shows, a purely objectivistic dialectics remains insufficient to Lukács since it is not yet for-itself.

7 I thus disagree with the eco-Marxist criticism that only sees dualism in Lukács's position in the *Defence* (see Burkett 2013).

8 To clarify this point of difference we might say that practical materialism helps us conduct better social science while critical materialism is a critique of social science in analogy with Marx's critique of political economy (see Gunn 1992). On Feenberg's interpretation of Lukács, see also Pellizzoni (2015: 193–6) and Westerman (2019: 242–60).

9   At some distance from the theoretical traditions I discuss here there are also attempts to deconstruct the dichotomy between nature and society, inspired by assemblage theory and actor-network theory (see Chapter 9). These attempts will not be discussed here since they are generally not driven by interest in dialectics or the Lukács problem (but for an exception, see Loftus 2012: 62–6).

10 Bloch's solution has been criticized for being conceptual mythology and lacking basis in Marx (e.g., Feenberg 2014: 50f, Jay 1984: 185f). Schmidt points out that anchoring dialectics in such a nature-subject leads to a dialectics surprisingly close to Engels's since such a dialectics would play itself out independently of humans and be possible to study in a contemplative mode, in the manner of traditional theory (Schmidt 2014: 210 n153).

11 For criticisms of Vogel's social constructivism, see also Cook (2011: 41), Malm (2018: 35ff), Soper (2009), and Feenberg (1999, 2014: 132–7). Malm makes the important point that the fact that nature untouched by human hands is disappearing does not entitle us to speak of nature as a whole as a human construction. That we affect nature doesn't mean that we construct it: "If I cut and mould wood into a bookcase, I have undoubtedly built that bookcase – but if I cut a branch off a tree, have I also built *that tree*?" (Malm 2018: 37). A modified constructivism is more defensible, Westerman tries to rescue Lukács by stressing that it is not the objectivity of natural science that is in question, but the capitalist forms for interacting with nature. Overcoming capitalism wouldn't affect the laws of nature, but might imply a different, non-coercive relation to the material world. "Lukács's interest in 'nature' is with the value-laden socially defined category, not with the material world as such" (Westerman 2019: 243). The problem with such a partial constructivism is that it doesn't provide any clues regarding the role of science when we relate to the objectivity of the material world. Here constellations are preferable, as I will argue below.

12 On constellations and montage in Benjamin, see Buck-Morss (1991: 67, 159).

13 This point also marks a point of contrast with Horkheimer, who, while insisting on a social contextualization of natural science in a manner that might seem analogous to what Adorno advocates, still idealistically believed in the ability of such a contextualization to correctly grasp its object (Breuer 1993).

14 It is no wonder that one of the best overviews in English of the Minamata disease, a chapter in Brett Walker's *The Toxic Archipelago* (2010), takes a form closely resembling a constellation, moving between natural history, early modern history, gender ideology, technology, myths, and protest activism.

Chapter 7

1. This chapter builds on and expands the arguments in Cassegård (2017).
2. This closeness is underlined by the fact that Western Marxists such as Bloch, Benjamin, Adorno, and Marcuse also adhere to a view of nature as alive and as a potential subject (see Feenberg 2014: 44ff, 160, 196, Flodin 2011, Loftus 2012: 25, 27, 38, 114).
3. While these scientists have expressed support for dialectical approaches as well as strong political commitments (e.g., Levins & Lewontin 1985, Lewontin & Levins 1998), it is hard to see how they methodologically break with a contemplative frame. The aspiration to bring together natural and social science also has to face the hurdle, pointed out by Feenberg, that it is only in the latter that dialectical inquiry brings about and is affected by an increasing self-awareness in the object (Feenberg 2014: 135ff).
4. As Cook points out: "Adorno's thoroughly dialectical view of natural history puts paid to Foster's contentious and largely unsupported claim that Western Marxists, including Adorno, 'increasingly rejected realism and materialism'" (2011: 25).

Chapter 8

1. See Foster (2016a, 2016b, 2018), Foster and Burkett (2018, Foster and Clark (2020, esp. chs 1 & 9), and Moore (2015: 75–87, 2016b, 2017). Other participants in the debate include Gellert (2019), Kincaid (2017), Malm (2018), Murphy and Angus (2016), Nayeri (2016), and Saito (2017b).
2. E.g. Arrighi (2004), Bonefeld (2002), Federici (2004), Fraser and Jaeggi (2018), Hardt and Negri (2009), Harvey (2005), Parenti (2015), Prudham (2013), and Sassen (2010).
3. For an explicit discussion of Marx's value law, see Moore (2014b). Moore is often misunderstood on this point, as for example when Foster and Burkett (2018) accuse Moore of trying to deconstruct the Marxian theory of value. In fact, Moore stands out from world-systems theory generally by his firm adherence to Marxian value theory. A contrasting approach is that of Hornborg (2001), who is similarly influenced by world-systems theory but differs in seeing flows of value and energy as separate. Unlike Moore, he sees no need to explain the unequal flows of use-value (appropriation) by connecting them to the value law.

The flows of use-value are said to benefit the world-system core by the transfer of productive potential per se rather than by their contribution to preventing the profit fall from falling. In Hornborg the connection between the two flows instead exists on the ideological level, since prices obscure the reality of the unequal exchange of use-values. See Moore (2000), which criticizes Hornborg for unnecessarily positing two different kinds of exploitation.

4   Building on Moore, Collard and Dempsey (2017) work out a typology of how humans as well as non-human labor relate to value production.

5   See the debate between Foster and Harvey in the *Monthly Review* 49(11) 1998 (Foster 1998, Harvey 1998; compare Harvey 1996: 146ff, 194ff), in which Harvey adopted a more constructivist, anti-Malthusian approach that criticized the idea of natural limits and the catastrophist rhetoric of Foster.

6   In terms of Castree's (2005) typology of positions regarding nature, Moore is thus closer to a "post-natural" thinking that challenges the society-nature dualism than to social constructivism.

7   In his attacks on Moore, Foster warns about the undue influence in Moore's writings from non-Marxist and even pro-capitalist approaches like that of Bruno Latour—who, Foster carefully points out, is a Senior Fellow at a leading center for capitalist-friendly eco-modernism, the Breakthrough Institute. In his attacks, Foster thus places Moore as link in a chain of influences including the Frankfurt School, first-generation eco-Marxists, the production-of-nature approach, and actor-network theory (Foster 2016a, 2016b, Foster & Clark 2016a, 2016b).

8   As Andreas Malm points out, dualism returns in Moore's writings when he contrasts the finite character of the biosphere and the infinite character of capital's demands (Malm 2018: 182).

Chapter 9

1   For overviews and presentations of new materialism, see Coole and Frost (2010). For object-oriented ontology, see Harman (2018). For a collection that brings together both approaches under the label "the nonhuman turn," see Grusin (2015). For critical discussions of some of these approaches, see Bennett (2015), Eagleton (2016: 10ff), Lemke (2017), Pellizzoni (2015), Malm (2018), Sheldon (2015), and Wolfe (2017). Much of the criticism of these approaches concern points (such as the agency of matter and the methodological ban on totalization)

that are also in focus for much criticism of actor-network theory (see e.g. Kirsch & Mitchell 2004, Lave 2015, Malm 2018, Sayes 2017, Söderberg 2017, Wilding 2010).
2. On actor-network theory and its criticism of macro-concepts, see Latour (2005).
3. The idea of a "dark ecology" is further developed in Morton (2016), where such ecology is said to be "dark" since it compels us to recognize melancholy, trauma, humiliations, and shocks. This work makes the political thrust of Morton's argument somewhat clearer through his argument that depression can develop into play. It remains unclear, however, how dark ecology will be effective against the catastrophic loops that he associates with what he calls agrilogistics.
4. In a recent defense of the production-of-nature approach, Ekers and Loftus (2012: 234) refer precisely to such examples as support.

Chapter 10

1. Thus, Smith has argued that the *Dialectic of Enlightenment* ends up in a political cul-de-sac since "if such domination is an inevitable aspect of social life, the only political alternatives are an anti-social (literally) politics of nature or else resignation to a kinder, gentler domination" (Smith 2007: 24). In similar vein, Wilding has argued that Adorno and Horkheimer fail to provide an account of practices that resist nature's destruction that could be building blocks for an environmental politics (Wilding 2008: 55). Biro (2000), seeing no room for political action in Adorno, argues for the need to turn to Marcuse, while Buck (2005) defends Adorno's negative, critical gaze on the present. Against these positions, I want to stress that Adorno's negative dialectics is not as negative as it appears and that utopian visions of reconciliation are a natural outgrowth of it.
2. Holloway's (2005: 150–3) thoughts on the scream of non-identity offers an idea of how an Adorno-inspired activism conjoined with critique might look.
3. For the debate on Habermas's anthropocentrism, see e.g. Whitebook (1979).
4. As Kohn (2013: 227) writes: "if 'we' are to survive the Anthropocene ... we will have to actively cultivate these ways of thinking with and like forests." A forest consists, not only of facts, but also of signs. Habermas might argue that the idea that trees can speak is anthropomorphism, an illusory projection. But even projections can challenge our thoughts and force us to think anew. What is crucial isn't objections voiced in language per se, but the reflection it evokes in the hearer.

5   This narrative has been promoted by Habermas as part of the justification for the linguistic turn through which he sought to provide critical theory with a new normative foundation and thereby rescue hope in rationality and progress (see Habermas 1984: 373–86, 1992: 333, 1994: 106, 116–19, 126f, Honneth 1979). See O'Kane (2015, 2018a) for a critique.

6   See e.g. Adorno (1997: 61–78), Adorno and Horkheimer (1997: 41, 245–55), Benjamin (1977b: 256f, 1997b: 103f), Bloch (1995: 669–74). Compare Alford (1985), and Cook (2011: 87, 110, 121–54).

7   See Chapter 3. For Adorno's idea that freedom will consist in "nature becoming conscious of itself," see Cook (2006). For the thesis that reason must come to insight into its own naturalness in order to break the spell of nature and the cycle of domination, see Adorno and Horkheimer (1997: xvi, 40) and Horkheimer (2013: 136f, 177, 179).

8   Marcuse expressed emphatic notions of a liberation of nature already in *One-dimensional Man* (Marcuse 2002: 240ff). For discussions of utopia and nature in Marcuse, see Biro (2000), Feenberg (2014: 196ff), and Vogel (2004: 242ff). For Marcuse's anticipation of various environmental and ecological movements, see Luke (1994).

9   An illustration of that impulse, it seems to me, is when Adorno (2005: 151), as an example of genuine progress, refers to the weeping Nietzsche forgetting his condemnation of pity and embracing a whipped horse in Turin.

References

Adorno, T. W. (1973a) *Negative Dialectics* (tr. E. B. Ashton), London: Routledge & Kegan Paul.

Adorno, T. W. (1973b) "Die Idee der Naturgeschichte", pp. 345–65, in R. Tiedemann (ed.) *Ges.Schriften Bd I, Philosophische Frühschriften*, Frankfurt/M: Suhrkamp.

Adorno, T. W. (1974) *Philosophische Terminologie*, Vol. 2 (ed. R. zur Lippe), Frankfurt/M: Suhrkamp.

Adorno, T. W. (1975) *Negative Dialektik*, Frankfurt/M: Suhrkamp.

Adorno, T. W. (1976) "Sociology and Empirical Research", pp. 68–86, in G. Adey & D. Frisby (tr.) *The Positivist Dispute in German Sociology*, London: Heinemann.

Adorno, T. W. (1978) *Minima Moralia* (tr. E. F. N. Jephcott), London: Verso.

Adorno, T. W. (1981) *Prisms* (tr. S. Weber & S. Weber), Cambridge, MA: MIT Press.

Adorno, T. W. (1987) "Late Capitalism or Industrial Society?", pp. 232–47, in V. Meja, D. Misgeld & N. Stehr (eds.) *Modern German Sociology*, New York: Columbia University Press.

Adorno, T. W. (1991) *Notes to Literature* (tr. S. Weber Nicholsen), Vol. I, New York: Columbia University Press.

Adorno, T. W. (1993) *Hegel: Three Studies* (tr. S. Weber Nicholsen), Cambridge, MA: MIT Press.

Adorno, T. W. (1997) *Aesthetic Theory* (tr. R. Hullot-Kentor), Minneapolis: University of Minnesota Press.

Adorno, T. W. (2005) *Critical Models: Interventions and Catchwords* (tr. H. Pickford), New York: Columbia University Press.

Adorno, T. W. (2008) *Lectures on Negative Dialectics: Fragments of a Lecture Course 1965/1966* (ed. R. Tiedemann, tr. R. Livingstone), Cambridge: Polity Press.

Adorno, T. W. (2018) "Theodor W. Adorno on 'Marx and the Basic Concepts of Sociological Theory': *From a Seminar Transcript in the Summer Semester of 1962*" (tr. V. Erlenbusch-Anderson & C. O'Kane), *Historical Materialism* 26(1): 154–64.

Adorno, T. W. & W. Benjamin (1999) *The Complete Correspondence 1928–1940* (ed. H. Lonitz, tr. N. Walker), Cambridge, MA: Harvard University Press.

Adorno, T. W. & E. Bloch (1988) "Something's Missing: A Discussion between Ernst Bloch and Theodor W. Adorno on the Contradictions of Utopian Longing", pp. 1–17, in J. Zipes & F. Mecklenburg (tr.) *The Utopian Function of Art and Literature*, Cambridge, MA: MIT Press.

Adorno, T. W. & M. Horkheimer (1997) *Dialectic of Enlightenment* (tr. J. Cumming), London: Verso.

Alford, C. F. (1985) *Science and the Revenge of Nature: Marcuse and Habermas*, Tampa: University Presses of Florida.

Allen, A. (2017) *The End of Progress; Decolonizing the Normative Foundations of Critical Theory*, New York: Columbia University Press.

Althusser, L. (1969) *For Marx* (tr. B. Brewster), Harmondsworth: Penguin.

Altvater, E. (1993) *The Future of the Market: An Essay on the Regulation of Money and Nature after the Collapse of "Actually Existing Socialism"*, London: Verso.

Anguelovski, I. & J. Martínez Alier (2014) "The 'Environmentalism of the Poor' Revisited: Territory and Place in Disconnected Glocal Struggles", *Ecological Economics* 102: 167–76.

Arato, A. (1972) "Lukacs' Theory of Reification", *Telos* 11: 25–66.

Arrighi, G. (2004) "Spatial and Other 'Fixes' of Historical Capitalism", *Journal of World-Systems Research* X(2): 527–39.

Arthur, C. J. (2002) *The New Dialectic and Marx's Capital*, Leiden: Brill.

Arthur, C. J. (2003) "The Problem of Use-Value for a Dialectic of Capital", pp. 131–49, in R. Albritton & J. Simoulidis (eds.) *New Dialectics and Political Economy*, Houndmills: Palgrave Macmillan.

Arthur, C. J. (2015) "Dialectic of the Value-Form", pp. 67–81, in D. Elson (ed.) *Value: The Representation of Labour in Capitalism*, London: Verso.

Balibar, E. & I. Wallerstein (1991) *Race, Nation, Class: Ambiguous Identities*, London: Verso.

Banaji, J. (2015) "From the Commodity to Capital: Hegel's Dialectic in Marx's *Capital*", pp. 14–45, in D. Elson (ed.) *Value: The Representation of Labour in Capitalism*, London: Verso.

Beiser, F. (2005) *Hegel*, New York: Routledge.

Bell, A. (2017) "Life in Ruins: Ecological Disaster and Adorno's Idea of Natural History", *Telos* 179: 188–94.

Bellofiore, R. & T. Redolfi Riva (2015) "The *Neue Marx-Lektüre*: Putting the Critique of Political Economy Back into the Critique of Society", *Radical Philosophy* 189: 24–36.

Bellofiore, R. & T. Redolfi Riva (2018) "Hans-Georg Backhaus: The Critique of Premonetary Theories of Value and the Perverted Forms of Economic Reality", pp. 386–401, in B. Best, W. Bonefeld & C. O'Kane (eds.) *The Sage Handbook of Frankfurt School Critical Theory*, Vol. 1, London: Sage.

Benjamin, W. (1977a) "Zentralpark", pp. 230–50, in S. Unseld (ed.) *Illuminationen, Ausgewählte Schriften 1*, Frankfurt/M: Suhrkamp.
Benjamin, W. (1977b) "Über den Begriff der Geschichte", pp. 251–61, in S. Unseld (ed.) *Illuminationen, Ausgewählte Schriften 1*, Frankfurt/M: Suhrkamp.
Benjamin, W. (1985) *The Origin of German Tragic Drama* (tr. J. Osborne), London: Verso.
Benjamin, W. (1988) "Über die Sprache überhaupt und über die Sprache des Menschen", pp. 9–26, in S. Unseld (ed.) *Angelus Novus, Ausgewählte Schriften I1*, Frankfurt/M: Suhrkamp.
Benjamin, W. (1996) *Walter Benjamin: Selected Writings, 1938–1940* (ed. M. Bullock & M. W. Jennings), Cambridge, MA: Harvard University Press.
Benjamin, W. (1997a) *Charles Baudelaire: A Lyric Poet in the Era of High Capitalism* (tr. H. Zohn), London: Verso.
Benjamin, W. (1997b) *One-Way Street*, London: Verso.
Benjamin, W. (1999) *The Arcades Project* (tr. H. Eiland & K. McLaughlin), Cambridge, MA: The Belknap Press of Harvard University Press.
Bennett, J. (2010) *Vibrant Matter: A Political Ecology of Things*, Durham: Duke University Press.
Bennett, J. (2015) "Systems and Things: On Vital Materialism and Object-Oriented Philosophy", pp. 223–40, in R. Grusin (ed.) *The Nonhuman Turn*, Minneapolis: University of Minnesota Press.
Benton, T. (2011) "Humanism = Speciesism?: Marx on Humans and Animals", pp. 99–120, in J. Sanbonmatsu (ed.) *Critical Theory and Animal Liberation*, Lanham: Rowman & Littlefield.
Berman, M. (1989) "Georg Lukács's Cosmic Chutzpah", pp. 137–52, in J. Marcus & Z. Tarr (eds.) *Georg Lukács: Theory, Culture, and Politics*, New Brunswick: Transaction Publishers.
Biro, A. (2000) "Denaturalizing Ecological Politics: 'Alienation from Nature' from Rousseau to Marcuse", PhD dissertation, York University, Toronto.
Biro, A. (ed.) (2011a) *Critical Ecologies: The Frankfurt School and Contemporary Environmental Crises*, Toronto: University of Toronto Press.
Biro, A. (2011b) "Ecological Crisis and the Culture Industry Thesis", pp. 229–54, in A. Biro (ed.) *Critical Ecologies: The Frankfurt School and Contemporary Environmental Crises*, Toronto: University of Toronto Press.
Bloch, E. (1995) *The Principle of Hope*, Vol. 2 (tr. N. Plaice, S. Plaice & P. Knight), Cambridge, MA: MIT Press.
Bloch, E. (1998) "The Anxiety of the Engineer", in *Literary Essays* (tr. A. Joron and others), Stanford: Stanford University Press.

Bonefeld, W. (2002) "History and Social Constitution: Primitive Accumulation Is Not Primitive", *The Commoner*, March 2002: 1–8.

Bonefeld, W. (2004) "On Postone's Courageous but Unsuccessful Attempt to Banish the Class Antagonism from the Critique of Political Economy", *Historical Materialism* 12(3): 103–24.

Bonefeld, W. (2014) *Critical Theory and the Critique of Political Economy: On Subversion and Negative Reason*, New York: Bloomsbury.

Breuer, S. (1993) "The Long Friendship: Theoretical Differences between Horkheimer and Adorno", pp. 57–79, in S. Benhabib, W. Bonss & J. McCole (eds.) *On Max Horkheimer: New Perspectives*, Cambridge, MA: MIT Press.

Buck, C. (2005) "Experience First! Adorno and Radical Environmental Thought", www.yorku.ca/cnsconf/present/buck_cns.doc (accessed 2020-09-24).

Buck, C. (2013) "The Utopian Content of Reification: Adorno's Critical Social Theory of Nature", pp. 127–48, in C. Archer, L. Ephraim & L. Maxwell (eds.) *Second Nature: Rethinking the Natural through Politics*, New York: Fordham University Press.

Buck-Morss, S. (1977) *The Origin of Negative Dialectics: Theodor W. Adorno, Walter Benjamin and the Frankfurt Institute*, Hassocks: The Harvester Press.

Buck-Morss, S. (1991) *The Dialectics of Seeing: Walter Benjamin and the Arcades Project*, Cambridge, MA & London, England: MIT Press.

Burkett, P. (1997) "Nature in Marx Reconsidered: A Silver Anniversary Assessment of Alfred Schmidt's *Concept of Nature in Marx*", *Organization & Environment* 10(2): 164–83.

Burkett, P. (1999) "Nature's 'Free Gifts' and the Ecological Significance of Value", *Capital & Class* 68: 89–110.

Burkett, P. (2013) "Lukács on Science: A New Act in the Tragedy", *Historical Materialism* 21(3): 3–15.

Burkett, P. (2014) *Marx and Nature: A Red and Green Perspective*, Chicago: Haymarket Books.

Cassegård, C. (2007) *Shock and Naturalization in Contemporary Japanese Literature*, Folkestone: Global Oriental.

Cassegård, C. (2017) "Eco-Marxism and the Critical Theory of Nature: Two Perspectives on Ecology and Dialectics", *Distinktion: Journal of Social Theory* 18(3): 314–32.

Cassegård, C. & H. Thörn (2018) "Toward a Postapocalyptic Environmentalism? Responses to Loss and Visions of the Future in Climate Activism", *Environment and Planning E: Nature and Space* 1(4): 561–78.

Cassegård, C., H. Thörn, L. Soneryd & Å. Wettergren (eds.) (2017) *The Climate Action in a Globalizing World: Comparative Perspectives on Environmental Movements in the Global North*, New York: Routledge.

Castells, M. (1996) *The Information Age; Economy, Society and Culture. Volume 1. The Rise of the Network Society*, Oxford: Blackwell.

Castree, N. (2005) *Nature*, London: Routledge.

Clark, B. & R. York (2005a) "Dialectical Nature: Reflections in Honor of the Twentieth Anniversary of Levins and Lewontin's The Dialectical Biologist", *Monthly Review* 57(1): 13–22.

Clark, B. & R. York (2005b) "Dialectical Materialism and Nature: An Alternative to Economism and Deep Ecology", *Organization & Environment* 18(3): 318–37.

Collard, R.-C. & J. Dempsey (2017) "Capitalist Natures in Five Orientations", *Capitalism Nature Socialism* 28(1): 78–97.

Cook, D. (2006) "Nature Becoming Conscious of Itself: Adorno on Self-Reflection", *Philosophy Today* 50(3): 296–306.

Cook, D. (2011) *Adorno on Nature*, Durham: Acumen.

Coole, D. & S. Frost (eds.) (2010) *New Materialisms: Ontology, Agency, and Politics*, Durham: Duke University Press.

Creaven, S. (2000) *Marxism and Realism: A Materialistic Application of Realism in the Social Sciences*, London: Routledge.

Cronon, W. (1992) *Nature's Metropolis: Chicago and the Great West*, New York: W. W. Norton.

Cronon, W. (1995) "The Trouble with Wilderness; or, Getting Back to the Wrong Nature", pp. 69–90, in W. Cronon (ed.) *Uncommon Ground: Rethinking the Human Place in Nature*, New York: W. W. Norton.

Cunha, D. (2015) "The Anthropocene as Fetishism", *Mediations* 28(2): 65–77.

DeLanda, M. (2016) *Assemblage Theory*, Edinburgh: Edinburgh University Press.

Descola, P. (2013) *Beyond Nature and Culture* (tr. J. Lloyd), Chicago: University of Chicago Press.

Dupuy, J.-P. (2007/2008) "Rational Choice before the Apocalypse", *Anthropoetics— The Journal of Generative Anthropology* 13(3); http://www.anthropoetics.ucla.edu/ap1303/1303dupuy.htm (accessed 2016-02-03).

Dupuy, J.-P. (2013) *The Mark of the Sacred* (tr. M. B. DeBevoise), Stanford: Stanford University Press.

Eagleton, T. (2016) *Materialism*, New Haven: Yale University Press.

Ekers, M. & A. Loftus (2012) "Revitalizing the Production of Nature Thesis: A Gramscian Turn?", *Progress in Human Geography* 37(2): 234–52.

Elbe, I. (2010) *Marx im Westen. Die neue Marx-Lektüre in der Bundesrepublik seit 1965*, Berlin: Akademie Verlag.

Elbe, I. (2013) "Between Marx, Marxism, and Marxisms—Ways of Reading Marx' Theory", *View Point Magazine*, October 21; https://www.scribd.com/document/177878784/Ingo-Elbe-Between-Marx-Marxism-and-Marxisms-Ways-of-Reading-Marxs-Theory (accessed 2020-09-02).

Elbe, I. (2018) "Helmut Reichelt and the New Reading of Marx", pp. 367–85, in B. Best, W. Bonefeld & C. O'Kane (eds.) *The Sage Handbook of Frankfurt School Critical Theory*, Vol. 1, London: Sage.

Elson, D. (2015) "The Value Theory of Labour", pp. 115–80, in D. Elson (ed.) *Value: The Representation of Labour in Capitalism*, London: Verso.

Elson, D. (ed.) (2015 [1979]) *Value: The Representation of Labour in Capitalism*, London: Verso.

Engels, F. (1890) "Engels to J. Bloch in Berlin" (tr. S. Hook); https://www.marxists.org/archive/marx/works/1890/letters/90_09_21a.htm (accessed 2018-09-16).

Engels, F. (1987a) "Anti-Dühring", pp. 5–309, in E. Burns (tr.) *Karl Marx & Frederick Engels: Collected Works*, Vol. 25, London: Lawrence & Wishart.

Engels, F. (1987b) "Dialectics of Nature", pp. 313–588, in C. Dutt (tr.) *Karl Marx & Frederick Engels: Collected Works*, Vol. 25, London: Lawrence & Wishart.

Engster, F. & O. Schlaudt (2018) "Alfred Sohn-Rethel: Real Abstraction and the Unity of Commodity-Form and Thought Form", pp. 284–301, in B. Best, W. Bonefeld & C. O'Kane (eds.) *The Sage Handbook of Frankfurt School Critical Theory*, Vol. 1, London: Sage.

Enquist Källgren, K. (2019) *María Zambrano's Ontology of Exile: Expressive Subjectivity*, Cham: Palgrave Macmillan.

Federici, S. (2004) *Caliban and the Witch: Women, the Body and Primitive Accumulation*, New York: Autonomedia.

Feenberg, A. (1995) "Marcuse or Habermas: Two Critiques of Technology", *Inquiry* 39: 45–70.

Feenberg, A. (1999) "A Fresh Look at Lukács: On Steven Vogel's *Against Nature*", *Rethinking Marxism* 11(4): 83–93.

Feenberg, A. (2014) *The Philosophy of Praxis: Marx, Lukács and the Frankfurt School*, London: Verso.

Feenberg, A. (2015) "Lukács's Theory of Reification and Contemporary Social Movements", *Rethinking Marxism* 27(4): 490–507.

Fine, R. (2001) *Political Investigations: Hegel, Marx, Arendt*, London: Routledge.

Finlayson, J. G. (2012) "The Artwork and the *Promesse du Bonheur* in Adorno", *European Journal of Philosophy* 23(3): 392–419.

Finlayson, J. G. (2014) "Hegel, Adorno and the Origins of Immanent Criticism", *British Journal for the History of Philosophy* 22(6): 1142–66.

Flodin, C. (2011) "The Wor(l)d of the Animal: Adorno on Art's Expression of Suffering", *Journal of Aesthetics & Culture* 3(1): 1–12.

Foster, J. B. (1998) "The Scale of Our Ecological Crisis", *Monthly Review* 49(11): 5–16.

Foster, J. B. (2000) *Marx's Ecology: Materialism and Nature*, New York: Monthly Review Press.

Foster, J. B. (2013) "Marx and the Rift in the Universal Metabolism of Nature", *Monthly Review* 65(7): 1–19.

Foster, J. B. (2016a) "Marxism in the Anthropocene: Dialectical Rifts on the Left", *International Critical Thought* 6(3): 393–421.

Foster, J. B. (2016b) "In Defense of Ecological Marxism: John Bellamy Foster Responds to a Critic" (interview by I. Angus), *Climate & Capitalism*, June 6; http://climateandcapitalism.com/2016/06/06/in-defense-of-ecological-marxism-john-bellamy-foster-responds-to-a-critic/ (accessed 2016-06-10).

Foster, J. B. (2018) "Marx, Value, and Nature", *Monthly Review* 70(3): 122–36.

Foster, J. B. & P. Burkett (2016) *Marx and the Earth: An Anti-Critique*, Leiden: Brill.

Foster, J. B. & P. Burkett (2018) "Value Isn't Everything", *Monthly Review* 79(6): 1–17.

Foster, J. B. & B. Clark (2016a) "Marx's Ecology and the Left"; *Monthly Review* 68(2): 1–25.

Foster, J. B. & B. Clark (2016b) "Marx's Universal Metabolism of Nature and the Frankfurt School: Dialectical Contradictions and Critical Synthesis", pp. 110–35, in J. S. Ormrod (ed.) *Changing Our Environment, Changing Ourselves: Nature, Labour, Knowledge and Alienation*, London: Palgrave Macmillan.

Foster, J. B. & B. Clark (2016c) "Marxism and the Dialectics of Ecology", *Monthly Review* 68(5): 1–17.

Foster, J. B. & B. Clark (2020) *The Robbery of Nature: Capitalism and the Ecological Rift*, New York: Monthly Review Press.

Foster, J. B., B. Clark & R. York (2010) *The Ecological Rift: Capitalism's War on the Earth*, New York: Monthly Review Press.

Frase, P. (2016) *Four Futures: Visions of the World after Capitalism*, London: Verso.

Fraser, N. (2014) "Behind Marx's Hidden Abode: For an Expanded Conception of Capitalism", *New Left Review* 86: 55–72.

Fraser, N. & R. Jaeggi (2018) *Capitalism: A Conversation in Critical Theory* (ed. B. Milstein), Cambridge: Polity Press.

Gandler, S. (2018) "Praxis, Nature, Labour", pp. 734–49, in B. Best, W. Bonefeld & C. O'Kane (eds.) *The Sage Handbook of Frankfurt School Critical Theory*, Vol. 2, London: Sage.

Gellert, P. K. (2019) "Bunker's Ecologically Unequal Exchange, Foster's Metabolic Rift, and Moore's World-Ecology: Distinctions with or without a Difference?", pp. 107–40, in R. S. Frey, P. K. Gellert & H. F. Dahms (eds.) *Ecologically Unequal Exchange*, London: Palgrave Macmillan.

George, T. S. (2001) *Minamata: Pollution and the Struggle for Democracy in Postwar Japan*, Cambridge, MA: Harvard University Asia Center.

Görg, C. (2011) "Societal Relationships with Nature: A Dialectical Approach to Environmental Politics", pp. 43–72, in A. Biro (ed.) *Critical Ecologies: The Frankfurt School and Contemporary Environmental Crises*, Toronto: University of Toronto Press.

Gramsci, A. (1971) *Selections from Prison Notebooks* (eds. Q. Hoare & G. Nowell-Smith), London: Lawrence & Wishart.

Grusin, R. (ed.) (2015) *The Nonhuman Turn*, Minneapolis: University of Minnesota Press.

Guha, R. (1999) *Environmentalism: A Global History*, New York: Longman.

Guha, R. & J. Martinez-Alier (1997) *Varieties of Environmentalism: Essays North and South*, London: Earthscan.

Gunderson, R. (2014) "The First-Generation Frankfurt School on the Animal Question: Foundations for a Normative Sociological Animal Studies", *Sociological Perspectives* 57(3): 285–300.

Gunn, R. (1992) "Against Historical Materialism: Marxism as First-Order Discourse", pp. 1–45, in W. Bonefeld, R. Gunn & J. Holloway (eds.) *Open Marxism* Vol. II, London: Pluto Press.

Gunster, S. (2011) "Fear and the Unknown: Nature, Culture, and the Limits of Reason", pp. 206–28, in A. Biro (ed.) *Critical Ecologies: The Frankfurt School and Contemporary Environmental Crises*, Toronto: University of Toronto Press.

Habermas, J. (1971) "Technology and Science as 'Ideology'", pp. 81–122, in J. J. Shapiro (tr.) *Toward a Rational Society: Student Protest Science, and Politics*, London: Heinemann.

Habermas, J. (1984) *The Theory of Communicative Action: Reason and the Rationalization of Society*, Vol. 1 (tr. T. McCarthy), Cambridge: Polity Press.

Habermas, J. (1992) *The Theory of Communicative Action: Lifeworld and System: A Critique of Functionalist Reason*, Vol 2 (tr. T. McCarthy), Cambridge: Polity Press.

Habermas, J. (1993) *Justification and Application* (tr. C. P. Cronin), Cambridge: Polity Press.

Habermas, J. (1994) *The Philosophical Discourse of Modernity* (tr. F. G. Lawrence), Cambridge: Polity Press.

Hailwood, S. (2012) "Alienations and Natures", *Environmental Politics* 21(6): 882–900.

Hall, T. (2011) "Reification, Materialism, and Praxis: Adorno's Critique of Lukács", *Telos* 155: 61–82.

Hamilton, C. (2010) *Requiem for a Species: Why We Resist the Truth about Climate Change*, London: Earthscan.
Hamilton, C. (2017) *Defiant Earth: The Fate of Humans in the Anthropocene*, Cambridge: Polity Press.
Hardt, M. & A. Negri (2000) *Empire*, Cambridge, MA: Harvard University Press.
Hardt, M. & A. Negri (2009) *Commonwealth*, Cambridge, MA: The Belknap Press of Harvard University Press.
Harman, G. (2018) *Object-Oriented Ontology: A New Theory of Everything*, London: Penguin Books.
Harvey, D. (1996) *Justice, Nature and the Geography of Difference*, Oxford: Blackwell.
Harvey, D. (1998) "Marxism, Metaphors, and Ecological Politics", *Monthly Review* 49(11): 17–31.
Harvey, D. (2005) *A Brief History of Neoliberalism*, Oxford: Oxford University Press.
Harvey, D. (2010) *A Companion to Marx's Capital*, London: Verso.
Hegel, G. W. F. (1969) *Science of Logic* (tr. A. V. Miller), London: Routledge.
Hegel, G. W. F. (1975) *Hegel's Logic: Being Part One of the Encyclopaedia of the Philosophical Sciences (1830)* (tr. W. Wallace), Oxford: The Clarendon Press.
Hegel, G. W. F. (1983) "Will", in L. Rauch (tr.) *The Philosophy of Spirit (Jena Lectures 1805-6)*; https://www.marxists.org/reference/archive/hegel/works/jl/ch01b.htm (accessed 2019-09-17).
Hegel, G. W. F. (2001) "The Philosophy of History" (tr. J. Sibree), Kitchener: Batoche books.
Heinrich, M. (2012) *An Introduction to the Three Volumes of Karl Marx' Capital* (tr. A. Locascio), New York: Monthly Review Press.
Heitmann, L. (2018) "Society as 'Totality': On the Negative-Dialectical Presentation of Capitalist Socialization", pp. 589–606, in B. Best, W. Bonefeld & C. O'Kane (eds.) *The Sage Handbook of Frankfurt School Critical Theory*, Vol. 2, London: Sage.
Heynen, N. C., M. Kaika & E. Swyngedouw (eds.) (2006) *In the Nature of Cities: Urban Political Ecology and the Politics of Urban Metabolism*, Oxon: Routledge.
Hilferding, R. (1981) *Finance Capital: A Study of the Latest Phase of Capitalist Development* (tr. M. Watnick & S. Gordon), London: Routledge.
Holloway, J. (2005) *Change the World without Taking Power*, London, Ann Arbor: Pluto Press.
Holt, J. P. (2009) *Karl Marx's Philosophy of Nature, Action and Society: A New Analysis*, Newcastle upon Tyne: Cambridge Scholars Publishing.
Honneth, A. (1979) "Communication and Reconciliation: Habermas' Critique of Adorno", *Telos* 39: 45–61.
Honneth, A. (2012) *Reification: A New Look at an Old Idea* (ed. M. Jay), New York: Oxford University Press.

Horkheimer, M. (1988) "Materialismus und Metaphysik", pp. 70–105, in A. Schmidt (ed.) *Gesammelte Schriften, Bd 3: Schriften 1931–1936*, Frankfurt/M: Fischer.

Horkheimer, M. (2002a) "Traditional and Critical Theory", pp. 188–243, in *Critical Theory: Selected Essays* (tr. M. J. O'Connell), New York: Continuum.

Horkheimer, M. (2002b) "The Latest Attack on Metaphysics", pp. 132–87, in M. J. O'Connell (tr.) *Critical Theory: Selected Essays*, New York: Continuum.

Horkheimer, M. (2013) *Eclipse of Reason*, Mansfield Centre, TV: Martino Publishing.

Horkheimer, M. & T. Adorno (2002) *Dialectic of Enlightenment: Philosophical Fragments* (tr. E. Jephcott), Stanford: Stanford University Press.

Horkheimer, M. & T. Adorno (2010) "Towards a New Manifesto?" (tr. R. Livingstone), *New Left Review* 65: 32–61.

Hornborg, A. (2001) *The Power of the Machine: Global Inequalities of Economy, Technology, and Environment*, Walnut Creek: Altamira Press.

Huber, M. T. (2017) "Value, Nature, and Labor: A Defense of Marx", *Capitalism Nature Socialism* 28(1): 39–52.

Hussain, A. (2015) "Misreading Marx's Theory of Value: Marx's Marginal Notes on Wagner", pp. 82–101, in D. Elson (ed.) *Value: The Representation of Labour in Capitalism*, London: Verso.

Jameson, F. (1981) *The Political Unconscious: Narrative as a Socially Symbolic Act*, Ithaca, NY: Cornell University Press.

Jameson, F. (1991) *Postmodernism, or, the Cultural Logic of Late Capitalism*, Durham: Duke University Press.

Jameson, F. (2003) "Future City", *New Left Review* 21: 65–79.

Jarvis, S. (2004) "Adorno, Marx, Materialism", pp. 79–100, in T. Huhn (ed.) *The Cambridge Companion to Adorno*, Cambridge: Cambridge University Press.

Jay, M. (1973) *The Dialectical Imagination: A History of the Frankfurt School and the Institute of Social Research 1923–1950*, Berkeley: University of California Press.

Jay, M. (1984) *Marxism and Totality: The Adventures of a Concept from Lukács to Habermas*, Berkeley: University of California Press.

Katz, C. (1995) "Under the Falling Sky: Apocalyptic Environmentalism and the Production of Nature", pp. 276–82, in A. Callari, S. Cullenberg & C. Biewener (eds.) *Marxism in the Postmodern Age: Confronting the New World Order*, New York: The Guilford Press.

Kincaid, J. (2005) "A Critique of Value-Form Marxism", *Historical Materialism* 13(2): 85–119.

Kincaid, J. (2017) "Value Theory and the Schism in Eco-Marxism", *Historical Materialism*, May 25; http://www.historicalmaterialism.org/blog/value-theory-and-schism-eco-marxism (accessed 2017-05-30).

Kingsnorth, P. & D. Hine (2017) "Uncivilisation", pp. 257–84, in P. Kingsnorth (ed.) *Confessions of a Recovering Environmentalist and Other Essays*, London: Faber & Faber.

Kirsch, S. & D. Mitchell (2004) "The Nature of Things: Dead Labor, Nonhuman Actors, and the Persistence of Marxism", *Antipode* 36(4): 687–705.

Klein, N. (2008 [2007]) *The Shock Doctrine: The Rise of Disaster Capitalism*, London: Penguin.

Kocyba, H. (2018) "Alfred Schmidt: On the Critique of Social Nature", pp. 302–16, in B. Best, W. Bonefeld & C. O'Kane (eds.) *The Sage Handbook of Frankfurt School Critical Theory*, Vol. 1, London: Sage.

Kohn, E. (2013) *How Forests Think: Toward an Anthropology beyond the Human*. Berkeley: University of California Press.

Koolhaas, R. (2002) "Junkspace", *October* 100: 175–90.

Lange, E. L. (2016) "The Critique of Political Economy and The New Dialectic. Hegel, Marx, and Christopher J. Arthur's 'Homology Thesis'", *Crisis and Critique* 3(3): 235–72.

Latour, B. (2005) *Reassembling the Social: An Introduction to Actor-Network-Theory*, Oxford: Oxford University Press.

Lave, R. (2015) "Reassembling the Structural: Political Ecology and Actor-Network Theory", pp. 213–23, in T. Perrault, J. McCarthy & G. Bridge (eds.) *Handbook of Political Ecology*, London: Routledge.

Lemke, T. (2017) "Materialism without Matter: the Recurrence of Subjectivism in Object-Oriented Ontology", *Distinktion: Journal of Social Theory* 18(2): 133–52.

Levins, R. & R. Lewontin (1985) *The Dialectical Biologist*, Cambridge, MA: Harvard University Press.

Lewontin, R. & R. Levins (1998) "How Different Are Natural and Social Science?", *Capitalism Nature Socialism* 9(1): 85–9.

Liedman, S.-E. (2018) *A World to Win: The Life and Works of Karl Marx* (tr. J. N. Skinner), London: Verso.

Lilley, S., D. McNally, E. Yuen & J. Davis (2012) *Catastrophism: The Apocalyptic Politics of Collapse and Rebirth*, Oakland: PM Press.

Loftus, A. (2012) *Everyday Environmentalism: Creating an Urban Political Ecology*, Minneapolis: University of Minnesota Press.

Lukács, G. (1971) *History and Class Consciousness: Studies in Marxist Dialectics* (tr. R. Livingstone), London: Merlin Press.

Lukács, G. (1996) *The Theory of the Novel* (tr. A. Bostock), Cambridge, MA: MIT Press.

Lukács, G. (2000) *A Defence of History and Class Consciousness: Tailism and the Dialectic* (tr. E. Leslie), London: Verso.

Lukács, G. (2009) *Lenin: A Study on the Unity of His Thought* (tr. N. Jacobs), London: Verso.

Luke, T. W. (1994) "Marcuse and Ecology", pp. 189–207, in J. Bokina & T. J. Lukes (eds.) *Marcuse: From the New Left to the Next Left*, Lawrence: University Press of Kansas.

Luxemburg, R. (2003) *The Accumulation of Capital* (tr. A. Schwartzschild), London: Routledge.

Malm, A. (2016) *Fossil Capital: The Rise of Steam Power and the Roots of Global Warming*, London: Verso.

Malm, A. (2018) *The Progress of This Storm: Nature and Society in a Warming World*, London: Verso.

Marcuse, H. (1972) *Counter-revolution and Revolt*, Boston: Beacon Press.

Marcuse, H. (2002) *One-dimensional Man: Studies in the Ideology of Advanced Industrial Society*, London: Routledge.

Martinez-Alier, J. (2003) *The Environmentalism of the Poor: A Study of Ecological Conflicts and Valuation*, Cheltenham: Edward Elgar Publishing.

Marx, K. (1858) "Marx to Ferdinand Lassalle in Düsseldorf", https://marxists.catbull.com/archive/marx/works/1858/letters/58_02_22.htm (accessed 2018-09-16).

Marx, K. (1868) "Marx to Kugelmann in Hannover", London, July 11, 1868; https://www.marxists.org/archive/marx/works/1868/letters/68_07_11-abs.htm (accessed 2019-11-28).

Marx, K. (1981) *Economic and Philosophic Manuscripts of 1844* (tr. M. Milligan), London: Lawrence & Wishart.

Marx, K. (1990) *Capital: A Critique of Political Economy: Volume 1* (tr. B. Fowkes), London: Penguin.

Marx, K. (1991) *Capital: A Critique of Political Economy: Volume 3* (tr. D. Fernbach), London: Penguin.

Marx, K. (1992) *Capital: A Critique of Political Economy: Volume 2* (tr. D. Fernbach), London: Penguin.

Marx, K. (1993) *Grundrisse: Foundations of the Critique of Political Economy* (tr. M. Nicolaus), London: Penguin.

Marx, K. (2010a) "Difference between the Democritean and Epicurean Philosophy of Nature", pp. 25–107, in *Marx Engels Collected Works*, Vol. 1, London: Lawrence & Wishart.

Marx, K. (2010b) "The Eighteenth Brumaire of Louis Bonaparte", pp. 99–197, in *Marx Engels Collected Works*, Vol. 11, London: Lawrence & Wishart.

Marx, K. (2010c) "Preface" [to *A Contribution to the Critique of Political Economy*], pp. 261–5, in *Marx Engels Collected Works*, Vol. 29, London: Lawrence & Wishart.

Marx, K. (2010d) "Critique of the Gotha Programme", pp. 75–99, in *Marx Engels Collected Works*, Vol. 24, London: Lawrence & Wishart.

Marx, K. & F. Engels (2010) "The German Ideology", pp. 19–581, in *Marx Engels Collected Works*, Vol. 5, London: Lawrence & Wishart.

McKibben, B. (1989) *The End of Nature*, New York: Anchor books.

McNally, D. (2004) "The Dual Form of Labour in Capitalist Society and the Struggle over Meaning: Comments on Postone", *Historical Materialism* 12(3): 189–208.

Mendieta, E. (2011) "Animal Is to Kantianism as Jew Is to Fascism: Adorno's Bestiary", pp. 147–62, in J. Sanbonmatsu (ed.) *Critical Theory and Animal Liberation*, Lanham: Rowman & Littlefield.

Merchant, C. (1990) *The Death of Nature: Women, Ecology, and the Scientific Revolution*, San Francisco: Harper & Row.

Mészáros, I. (1970) *Marx's Theory of Alienation*, London: Merlin Press.

Methmann, C. P. & D. Rothe (2012) "Politics for the Day after Tomorrow: The Logic of Apocalypse in Global Climate", *Security Dialogue* 43(4): 323–44.

Moore, J. W. (2000) "Marx and the Historical Ecology of Capital Accumulation on a World Scale: A Comment on Alf Hornborg's 'Ecosystems and World Systems: Accumulation as an Ecological Process'". *Journal of World-Systems Research* 6(1): 133–8.

Moore, J. W. (2014a) "Toward a Singular Metabolism", *New Geographies* 6: 10–19.

Moore, J. W. (2014b) "The Value of Everything? Work, Capital, and Historical Nature in the Capitalist World-Ecology", *Review* 37(3–4): 245–92.

Moore, J. W. (2015) *Capitalism in the Web of Life: Ecology and the Accumulation of Capital*, London: Verso.

Moore, J. W. (ed.) (2016a) *Anthropocene or Capitalocene? Nature, History, and the Crisis of Capitalism*, Oakland: PM Press.

Moore, J. W. (2016b) "Metabolisms, Marxisms, & Other Mindfields", posted on October 16, *World-Ecological Imaginations: Power and Production in the Web of Life* (blog); https://jasonwmoore.wordpress.com/ (accessed 2020-02-22).

Moore, J. W. (2016c) "Nature/Society & the Violence of Real Abstraction", posted on October 4, *World-Ecological Imaginations: Power and Production in the Web of Life* (blog); https://jasonwmoore.wordpress.com/ (accessed 2020-02-22).

Moore, J. W. (2017) "Metabolic Rift or Metabolic Shift? Dialectics, Nature, and the World-Historical Method", *Theory and Society* 46(4): 285–318.

Morton, T. (2007) *Ecology without Nature: Rethinking Environmental Aesthetics*, Cambridge, MA: Harvard University Press.

Morton, T. (2013) *Hyperobjects: Philosophy and Ecology after the End of the World*, Minneapolis: University of Minnesota Press.

Morton, T. (2016) *Dark Ecology: For a Logic of Future Coexistence*, New York: Columbia University Press.

Morton, T. (2017) *Humankind: Solidarity with Nonhuman People*, London: Verso.

Murphy, F. & I. Angus (2016) "Two Views on Marxist Ecology and Jason W. Moore", *Climate and Capitalism*, posted on June 23; http://climateandcapitalism.com/2016/06/23/two-views-on-marxist-ecology-and-jason-w-moore/ (accessed 2018-03-05).

Nayeri, K. (2016) "'Capitalism in the Web of Life'—A Critique", *Climate and Capitalism*, posted on July 19; http://climateandcapitalism.com/2016/07/19/capitalism-in-the-web-of-life-a-critique/ (accessed 2018-03-05).

O'Kane, C. (2015) "The Process of Domination Spews Out Tatters of Subjugated Nature: Critical Theory, Negative Totality and the State of Extraction", pp. 190–206, in The Black Box Collective (ed.) *Black Box: A Record of the Catastrophe*, Vol. 1, Oakland: PM Press.

O'Kane, C. (2018a) "'Society Maintains Itself Despite All the Catastrophes That May Eventuate': Critical Theory, Negative Totality, Authoritarianism and Crisis", *Constellations* 25(2): 287–301.

O'Kane, C. (2018b) "Moishe Postone's New Reading of Marx: the Critique of Political Economy as a Critical Theory of the Historically Specific Social Form of Labor", *Consecutio Rerum* 3(5): 485–501.

Ollman, B. (2003) *Dance of the Dialectic: Steps in Marx's Method*, Urbana: University of Illinois Press.

Parenti, C. (2015) "The Environment Making State: Territory, Nature, and Value", *Antipode* 47(4): 829–48.

Parson, S. (2017) "Cthulhuscene: Ecological Catastrophe, Cosmic Horror, and the Politics of Doom", *Reading Super Heroes Politically* (blog), February 24; https://readingsuperheroespolitically.wordpress.com/2017/02/24/cthulhuscene-ecological-catastrophe-cosmic-horror-and-the-politics-of-doom/ (accessed 2017-11-25).

Pellizzoni, L. (2015) *Ontological Politics in a Disposable World: The New Mastery of Nature*, Farnham: Ashgate.

Pensky, M. (2005) "Natural History: The Life and Afterlife of a Concept in Adorno", *Critical Horizons* 2(5): 227–58.

Postone, M. (1993) *Time, Labor, and Social Domination: A Reinterpretation of Marx's Critical Theory*, Cambridge: Cambridge University Press.

Postone, M. (2009) *History and Heteronomy: Critical Essays*, Tokyo: UTCP.

Prudham, S. (2013) "Men and Things: Karl Polanyi, Primitive Accumulation, and Their Relevance to a Radical Green Political Economy", *Environment and Planning A* 45: 1569–87.

Rabinbach, A. (1992) *The Human Motor: Energy, Fatigue, and the Origins of Modernity*, Berkeley: University of California Press.

Radkau, J. (2008) *Nature and Power: A Global History of the Environment* (tr. T. Dunlap), Cambridge: Cambridge University Press.

Rees, J. W. (1998) *The Algebra of Revolution: The Dialectic and the Classical Marxist Tradition*, London: Routledge.

Reichelt, H. (1982) "From the Frankfurt School to Value-Form Analysis", *Thesis Eleven* 4: 166–9.

Reichelt, H. (2007) "Marx's Critique of Economic Categories: Reflections on the Problem of Validity in the Dialectical Method of Presentation in *Capital*", *Historical Materialism* 15: 3–52.

Ricoeur, P. (1976) "Ideology and Utopia as Cultural Imagination", *Philosophical Exchange* 7(1): 17–28.

Rose, G. (1976) "How Is Critical Theory Possible? Theodor W. Adorno and Concept Formation in Sociology", *Political Studies* 24(1): 69–85.

Rose, G. (1978) *The Melancholy Science: An Introduction to the Thought of Theodor W. Adorno*, London: Macmillan Press.

Rubin, I. I. (1973) *Essays on Marx's Theory of Value* (tr. M. Samardzija & F. Perlman), Montreal: Black Rose Books.

Saito, K. (2017a) *Karl Marx's Ecosocialism: Capital, Nature, and the Unfinished Critique of Political Economy*, New York: Monthly Review Press.

Saito, K. (2017b) "Marx in the Anthropocene: Value, Metabolic Rift, and the Non-Cartesian Dualism", *Zeitschrift für kritische Sozialtheorie und Philosophie* 4(1–2): 276–95

Sanbonmatsu, J. (ed.) (2011a) *Critical Theory and Animal Liberation*, Lanham: Rowman & Littlefield.

Sanbonmatsu, J. (2011b) "Introduction", pp. 1–34, in J. Sanbonmatsu (ed.) *Critical Theory and Animal Liberation*, Lanham: Rowman & Littlefield.

Sassen, S. (2010) "A Savage Sorting of Winners and Losers: Contemporary Versions of Primitive Accumulation", *Globalizations* 7(1–2): 23–50.

Sayes, E. (2017) "Marx and the Critique of Actor-Network Theory: Mediation, Translation, and Explanation", *Distinktion: Journal of Social Theory* 18(3): 294–313.

Schmidt, A. (1968a) "Zum Erkenntnisbegripp der Kritik der politischen Ökonomie", pp. 30–43, in W. Euchner & A. Schmidt (eds.) (1968) *Kritik der politischen Ökonomie heute. 100 Jahre "Kapital"*, Frankfurt: Europäische Verlagsanstalt.

Schmidt, A. (1968b) "Diskussion", pp. 25–9, in W. Euchner & A. Schmidt (eds.) (1968) *Kritik der politischen Ökonomie heute. 100 Jahre "Kapital"*, Frankfurt: Europäische Verlagsanstalt.

Schmidt, A. (1973) *Emanzipatorische Sinnlichkeit: Ludwig Feuerbachs anthropologischer Materialismus*, München: Carl Hanser Verlag.

Schmidt, A. (1976) *Die Kritische Theorie as Geschichtsphilosophie*, München, Wien: Carl Hanser Verlag.

Schmidt, A. (1981) *History and Structure: An Essay on Hegelian-Marxist and Structural Theories of History* (tr. J. Herf), Cambridge, MA: MIT Press.

Schmidt, A. (1983) "Begriff der Materialismus bei Adorno", pp. 14–31, in L. v. Friedeburg & J. Habermas (eds.) *Adorno-Konferenz 1983*, Frankfurt/M: Suhrkamp.

Schmidt, A. (1986) "Aufklärung und Mythos im Werk Max Horkheimers", pp. 180–243, in A. Schmidt & N. Altwicker (eds.) *Max Horkheimer heute: Werk und Wirkung*, Frankfurt/M: Fischer.

Schmidt, A. (2014) *The Concept of Nature in Marx* (tr. B. Fowkes), London: Verso.

Schmidt, A. (2016) "Vorwort zur Neuauflage 1993. Für einen ökologischen Materialismus", preface to *Der Begripp der Natur in der Lehre von Marx*, Hamburg: CEP Europäische Verlagsantalt GmbH.

Sheldon, R. (2015) "Form / Matter / Chora: Object-Oriented Ontology and Feminist New Materialism", pp. 193–222, in R. Grusin (ed.) *The Nonhuman Turn*, Minneapolis: University of Minnesota Press.

Sloterdijk, P. (1987) *Critique of Cynical Reason* (tr. M. Eldred), London, Minneapolis: University of Minnesota Press.

Smith, N. (2006) "Foreword", pp. xi–xv, in N. C. Heynen, M. Kaika & E. Swyngedouw (eds.) *In the Nature of Cities: Urban Political Ecology and the Politics of Urban Metabolism*, Oxon: Routledge.

Smith, N. (2007) "Nature as Accumulation Strategy", *Socialist Register* 43: 16–36.

Smith, N. (2010) *Uneven Development: Nature, Capital and the Production of Space*, London: Verso.

Smith, N. & P. O'Keefe (1980) "Geography, Marx and the Concept of Nature", *Antipode* 12(2): 30–9.

Söderberg, J. (2017) "The Genealogy of 'Empirical Post-structuralist' STS, Retold in Two Conjunctures: The Legacy of Hegel and Althusser", *Science as Culture* 26(2): 185–208.

Sohn-Rethel, A. (1978) *Intellectual and Manual Labor*, Atlantic Highlands, NJ: Humanities Press.

Soper, K. (1995) *What is Nature? Culture, Politics and the Non-Human*, Oxford: Blackwell.

Soper, K. (2009) "Unnatural Times? The Social Imaginary and the Future of Nature", *Sociological Review* 57(2): 222–35.

Stoner, A. M. & A. Melathopoulos (2015) *Freedom in the Anthropocene: Twentieth-century Helplessness in the Face of Climate Change*, New York: Palgrave Macmillan.

Swyngedouw, E. (2010) "Apocalypse Forever? Post-political Populism and the Spectre of Climate Change", *Theory Culture & Society* 27(2-3): 213–32.

Swyngedouw, E. (2013) "Apocalypse Now! Fear and Doomsday Pleasures", *Capitalism Nature Socialism* 24(1): 9–18.

Swyngedouw, E. (2015) "Urbanization and Environmental Futures: Politicizing Urban Political Ecologies", pp. 609–19, in T. Perrault, J. McCarthy & G. Bridge (eds.) *Handbook of Political Ecology*, London: Routledge.

Testa, I. (2007) "Criticism from within Nature: The Dialectic between First and Second Nature", *Philosophy & Social Criticism* 33(4): 473–97.

Thörn, H. (1997) *Rörelser i det moderna: Politik, modernitet och kollektiv identitet i Europa 1789-1989*, Stockholm: Tiden Athena.

Timpanaro, S. (1974) "Considerations on Materialism", *New Left Review* I 85: 3–22.

Toscano, A. (2008) "The Open Secret of Real Abstraction", *Rethinking Marxism* 20(2): 273–87.

Toscano, A. (2014) "Materialism without Matter: Abstraction, Absence and Social Form", *Textual Practice* 28(7): 1221–40.

Trenkle, N. (2014) "Value and Crisis: Basic Questions", pp. 1–15, in N. Larsen, M. Nilges, J. Robinson & N. Brown (eds.) *Marxism and the Critique of Value*, Chicago: MCM Publishing.

Truskolaski, S. (2014) "Adorno's Imageless Materialism", *Studies in Social and Political Thought* 23: 14–23.

Tsing, A. (2015) *The Mushroom at the End of the World: On the Possibility of Life in Capitalist Ruins*, Princeton: Princeton University Press.

Urry, J. (2011) *Climate Change and Society*, Cambridge: Polity.

van Reijen, W. & J. Bransen (2002) "The Disappearance of Class History in 'Dialectic of Enlightenment': A Commentary on the Textual Variants (1947 and 1944)", pp. 248–53, in M. Horkheimer & T. Adorno (eds.) *Dialectic of Enlightenment: Philosophical Fragments*, Stanford: Stanford University Press.

Vogel, S. (1996) *Against Nature: The Concept of Nature in Critical Theory*, Albany: State University of New York Press.

Vogel, S. (2004) "Marcuse and the 'New Science'", pp. 240–6, in J. Abromeit & W. M. Cobb (eds.) *Herbert Marcuse: A Critical Reader*, London: Routledge.

Vogel, S. (2011) "On Nature and Alienation", pp. 187–205, in A. Biro (ed.) *Critical Ecologies: The Frankfurt School and Contemporary Environmental Crises*, Toronto: University of Toronto Press.

Vogel, S. (2015) *Thinking like a Mall: Environmental Philosophy after the End of Nature*, Cambridge, MA: MIT Press.

Wainwright, J. & G. Mann (2018) *Climate Leviathan: A Political Theory of Our Planetary Future*, London: Verso.

Walker, B. L. (2010) *Toxic Archipelago: A History of Industrial Disease in Japan*, Seattle: University of Washington Press.

Walker, R. (2017) "Value and Nature: Rethinking Capitalist Exploitation and Expansion", *Capitalism Nature Socialism* 28(1): 53–61.

Weber Nicholsen, S. (2002) *The Love of Nature and the End of the World: The Unspoken Dimension of Environmental Concern*, Cambridge: MIT Press.

Westerman, R. (2019) *Lukács's Phenomenology of Capitalism: Reification Revalued*, Cham: Springer Nature / Palgrave Macmillan.

Whitebook, J. (1979) "The Problem of Nature in Habermas", *Telos* 40: 41–69.

Wiggershaus, R. (1994) *The Frankfurt School: Its History, Theories and Political Significance* (tr. M. Robertson), Cambridge: Polity Press.

Wilding, A. (2008) "Ideas for a Critical Theory of Nature", *Capitalism, Nature, Socialism*, 19(4): 48–67.

Wilding, A. (2010) "Naturphilosophie Redivivus: On Bruno Latour's 'Political Ecology'", *Cosmos and History: The Journal of Natural and Social Philosophy* 6(1): 18–32.

Wilke, S. (2016) "Enlightenment, Dialectic, and the Anthropocene: Bruised Nature and the Residues of Freedom", *Telos* 177: 83–106.

Williams, E. C. (2011) *Combined and Uneven Apocalypse*, Winchester: Zero Books.

Williams, R. (1978) "Problems of Materialism", *New Left Review* I: 109: 3–17.

Williams, R. (1980) *Problems in Materialism and Culture: Selected Essays*, London: NLB.

Williams, R. (1983) *Keywords: A Vocabulary of Culture and Society*, New York: Oxford University Press.

Wolfe, C. T. (2017) "Materialism New and Old", *Antropología Experimental* 17: 215–24.

Worster, D. (1994) *Nature's Economy: A History of Ecological Ideas*, Cambridge: Cambridge University Press.

Yates, M. (2011) "Toward a Green Marxist Cultural Studies: Notes on Labor, Nature, and the Historical Specificity of Capitalism", pp. 238–44, in P. Smith (ed.) *The Renewal of Cultural Studies*, Philadelphia: Temple University Press.

Yates, M. (2018) "Environmentalism and the Domination of Nature", pp. 1503–19, in B. Best, W. Bonefeld & C. O'Kane (eds.) *The Sage Handbook of Frankfurt School Critical Theory*, Vol. 3, London: Sage.

Žižek, S. (2008) *The Ticklish Subject: The Absent Centre of Political Ontology*, London: Verso.

Index

abstract labor. *See under* labor
actor-network theory 160–1, 169, 172–3, 179
Adorno, T. W. 5, 7–8, 10–11, 13–14, 17, 22–4, 44, 49–77, 79, 101, 112, 148–9, 151, 163–5, 171–2, 174, 176, 183, 187–9, 191, 194–5, 197–8, 200–1, 203–5
    constellations 124–8, 131–2, 135
    critical materialism 38–40
    and Schmidt 79, 83–4, 89, 92–3, 95–7
Althusser, L. 31, 141
animals 49, 57, 105, 191, 201
Anthropocene 7–9, 190, 192, 199
apocalypse 15, 188, 198, 200. *See also* post-apocalypse
appropriation 154–5, 165, 167
Arthur, C. 34

Benjamin, W. 8, 13–14, 59, 63, 72–3, 90, 124–5, 149, 164, 175, 188, 195, 197, 200
Bennett, J. 9, 24–5, 169, 171–3, 178–9, 182, 184–5, 191
Biro, A. 17
Bloch, E. 1–2, 5–6, 84, 122, 131, 189, 204
Bonefeld, W. 42
Buck, C. 173
Buck-Morss, S. 60
Burkett, P. 18, 87–91, 104, 137, 152

Capitalocene 199
Cartesian dualism. *See* dualism
Castree, N. 18
catastrophe vi, 4, 67–8, 70, 79, 160, 177, 183, 190, 192, 197, 200, 203–4. *See also* permanent catastrophe
causal materialism 11, 27–32, 44–6, 57–8, 83, 122–3, 128, 147, 151, 198
class 21, 74–6, 111, 157, 193
climate change. *See* global warming

commodity fetishism 8, 101–2, 177–8
communication 191–2
concrete labor. *See under* labor
constellations 24, 26, 44, 56, 122, 124–35, 149–50, 188–9, 198, 204–5
constructivism 16–20, 37, 51, 61–2, 64, 122–3, 131, 156, 163–4, 219 n.11
critical materialism 11–13, 22, 27–8, 38–47, 50, 55–8, 73, 83–4, 95, 98, 110, 122, 130, 151, 153, 166–7, 171, 187–8, 198
critical subject 20–1, 25, 47, 187–9, 200–2
critical theory of nature 1, 7, 13–20, 27, 204
critique vii
    immanent 12, 19, 28, 39, 44, 52–3, 76, 96, 132, 165, 187, 198–9
    interplay with catastrophes 14–16, 21, 200, 202–3
    of political economy 5, 12, 39–40, 58, 100, 109, *see also* critical subject
Cunha, D. 7

Dark Mountain 70, 200
Dean, J. 187
Dialectic of Enlightenment 50–3, 63–6, 68–9, 71, 74–5, 91, 183, 189, 194
dialectics 11, 24–6, 29–30, 32–4, 36–7, 40–1, 82, 89–90, 118–19, 121–2, 130, 138, 148–51, 161
    of nature 11, 24, 30–1, 82, 123, 128, 137–49
dualism (of nature and society) 19–21, 25, 89, 119–22, 146, 153, 158–67, 180–5
Dupuy, J.-P. 69–70, 181

eco-Marxism 18–19, 23–4, 32, 87, 137–8, 150
Engels, F. 1, 11, 28–31, 33–4, 36, 80, 82–3, 85, 89, 94, 123, 134, 137, 140–1, 143

environmentalism 15, 20, 70–1, 73, 175, 184, 190, 193, 195, 198, 200
Epicurus 138, 140–2
experience, reflection on 52–3, 132, 191–2

Feenberg, A. 121–3, 128, 130–1
Feuerbach, L. 93–4, 195
form and matter 5–6, 12, 16, 19, 21, 23, 153, 162–7, 170, 183, 187–8, 190, 204, 206 n. 1
Foster, J. B. 18, 24, 32, 89–90, 92, 109, 117, 122–3, 134, 137–153, 158, 160–3, 165, 167
Fraser, N. 42, 112–14, 193

gender 112, 157
global warming vi, 3–4, 54, 69, 126, 158, 177, 183, 190, 203
Görg, C. 181, 183
Gramsci, A. 11, 32, 34

Habermas, J. 191–2
Hamilton, C. 7
Harvey, D. 18, 154
Hegel, G. W. F. 2–5, 12–14, 29–31, 54–5, 84, 95–6, 110, 126, 183, 200–1, 206 n. 1
Hilferding, R. 29
historization 71–2, 193
Horkheimer, M. 6–7, 38, 49–50, 166, 187, 201–3

idealism 4, 12–14, 29–30, 39, 55–8, 64, 67–8, 75–6, 101, 203
ideology 61, 71–2, 199
immanent critique. *See under* critique
instrumental reason 49, 51, 60–1, 65, 90, 92–4, 181–2

Jameson, F. 12, 38, 202
Jay, M. 74

Koolhaas, R. 202

labor 33, 35, 91, 93–4, 99–106, 110–12
    abstract and concrete 84, 91, 94–5, 97–8, 100–2, 105, 107, 109–11, 114–15
Lucretius 140, 142–3

Lukács, G. 7–8, 11, 32–4, 37–8, 58, 60, 62–3, 84, 118–23, 129–31, 137, 144, 147–8, 151, 163, 217 n. 4
Lukács problem, the 22, 24, 31, 37–8, 43, 47, 118–24, 131, 134, 149–50, 198, 204
Luxemburg, R. 154, 157

McKibben, B. 71
Malm, A. 123
Marcuse, H. 10, 117, 149, 195, 197
Marx, K. 5–10, 12, 18, 27–42, 76, 79–90, 93–7, 99–115, 117, 120, 125, 133, 137, 139–42, 144, 150, 154, 159, 170, 177, 182, 188, 193, 196
materialism 27–47, 139–44, 147–8. *See also* causal materialism, critical materialism, and practical materialism
Melathopoulos, A. vi
Mészáros, I. 36, 150
metabolic rift 18, 92, 109, 138, 158, 160
metabolism 18, 80–1, 84–6, 91–2, 95, 97–9, 107–10, 163–4, 196
mimesis 26, 53, 57, 93, 149, 171, 173, 195, 201–3
Minamata 55, 117, 126–7
monism 153, 158, 160–2, 165–7, 170, 174
Moore, J. 24–5, 108, 153–67, 170, 182, 193
Morton, T. 1, 24–5, 169–71, 174–80, 184–5

natural beauty 71–2
natural history 17, 22, 35–6, 49–51, 59–73, 75, 163–5, 176, 192–3
natural science 18–19, 24, 30–1, 34–7, 46, 82, 108, 117–24, 127–30, 133–4, 138–9, 144–5, 149–50
naturalization 72–3, 192–3
nature
    concept of vi, 16–18, 35, 49–50, 59–60, 80, 88–9
    domination of 10, 19, 23, 51, 60–1, 64–7, 71, 74–5, 81, 90–2, 94, 97, 111, 158, 162, 166–7, 169, 181–3, 194–6
    inner 57, 64, 71, 189
    reconciliation with 25, 34, 47, 52–4, 80, 84, 86, 92, 97–8, 149, 175, 191–2, 194–7, 204
    struggle with 79, 84–8, 90, 93, 195, *see also* second nature

nature-society divide 20, 156, 162, 165–6, 169–70, 174, 181
negative dialectics 25–6, 44, 52–4, 76, 132, 188–9, 192, 204
negative ontology 81, 85
negative totality 12, 22–3, 40–4, 95–7, 100, 109, 133, 185, 187–8, 198
new Marx-reading, the 38, 79, 95–6
new materialism 160, 169, 171, 173
non-identity 13, 19, 21, 24–5, 39, 44–7, 50, 52–7, 63, 75–6, 83–4, 88, 92–3, 100, 107, 110–12, 126–7, 130, 133–5, 163–6, 170, 172, 174–6, 178–9, 183–5, 187–90, 203–5

object-oriented ontology 169
Ollman, B. 43
outside, problem of the 22, 24, 28, 40–2, 47, 97–8, 132–3, 135, 204

Parson, S. 70
patriarchy 41, 112
Pellizzoni, L. 181, 183
permanent catastrophe 13–16, 67, 188, 200–1
Plekhanov, G. 141
post-apocalypse 69–70, 131, 175
posthumanism 199
Postone, M. 106
practical materialism 11–12, 27–8, 32–8, 44–6, 58, 81, 83–4, 122–3, 129–30, 145, 147, 151, 187, 198
praxis 11, 25, 33–4, 36, 47, 82, 187–90, 200–1
    natural 24, 141–7
primacy of the object 17, 22, 49–50, 52, 54–8, 60–1, 63–4, 68, 73, 81, 83–4, 123, 125, 131, 149, 163–5, 172, 184, 200
primitive accumulation 42, 113, 154–5, 193
production-of-nature approach 18–19, 158–60, 162–4
profit rate 155–7

racism 112, 157, 193, 200
real abstraction 6, 43–4, 101, 133, 156, 167, 185
realism 17–19, 51, 56, 61–2
reification 19, 33, 37, 39, 58, 60, 62–4, 82, 84, 92, 101, 107–8, 118–19, 121, 129–30, 164, 166, 171, 173, 175, 178, 189, 196, 206 n. 4
    definition 8–10
Ricoeur, P. 198

Saito, Kohei 165
Schmidt, A. 10, 23, 36, 51, 79–99, 104, 107, 111, 114, 122, 150, 158–9, 163–4, 194–6
second nature 7–10, 19–21, 60, 62–3, 65, 69, 73, 80–1, 83, 95–6, 101, 107–9, 159, 164, 173, 184, 192–3, 199
self-preservation 64–5, 201–3
sexism 112, 193, 200
Smith, N. 18–19, 158–60, 162–4, 166–7, 181–2
social constructivism. See constructivism
social movements 14–15, 70, 73, 190, 203
socialism 34, 79, 81, 83–5, 88, 91–2, 120–1, 190, 195–7
Sohn-Rethel, A. 101, 156
Soper, K. 17, 61, 159
state power 41, 113, 206 n. 2
Stoner, A. vi
suffering 12–13

Timpanaro, S. 31
totality 11, 29–30, 33–4, 110–11, 118, 130, 179–80, 184–5, see also negative totality
Tsing, A. 130–1

use-value 80–1, 89, 98–103, 106–11, 113–15, 155, 159, 163–4, 167, 182, 199
utopia 36, 52, 61, 86–7, 149, 188, 190, 194, 197–9, 205
    problem of 22–3, 25, 47, 80, 97–9, 198, 204
    tool for critique 197–8, 205

value 23, 26, 76, 89, 91, 97–108, 111, 113–15, 167, 199
    contribution of nature to 99, 102–6, 154–5
    labor theory of 100–6, 155
    and material wealth 89, 99, 101, 106–7, 109

value-form 6, 18, 100–2, 114
Vogel, S. 37, 117–18, 122–3, 131, 134, 189

Weber, M. 92
Wiggershaus, R. 60
wilderness 70–2, 193
Wilding, A. 74

Williams, E. C. 15, 204
Williams, R. 17

Yates, M. 110
Yuen, E. 200

Žižek, S. 69

www.ingramcontent.com/pod-product-compliance
Lightning Source LLC
Chambersburg PA
CBHW052113010526
44111CB00036B/1964